Polymer Blends and Compatibilization

Special Issue Editor
Volker Altstädt

MDPI

Guest Editor
Volker Altstädt
Universität Bayreuth
Germany

Editorial Office
MDPI AG
St. Alban-Anlage 66
Basel, Switzerland

This edition is a reprint of the Special Issue published online in the open access journal *Materials* (ISSN 1996-1944) in 2016 (available at: http://www.mdpi.com/journal/materials/special_issues/Blends_Compatibilization).

For citation purposes, cite each article independently as indicated on the article page online and as indicated below:

Author 1; Author 2; Author 3 etc. Article title. *Journal Name*. **Year**. Article number/page range.

ISBN 978-3-03842-364-5 (Pbk)
ISBN 978-3-03842-365-2 (PDF)

Table of Contents

About the Guest Editor

Volker Altstädt is Full Professor of the Department of Polymer Engineering in the Faculty of Engineering (ING.) at the University of Bayreuth. Furthermore, since 2009 he is CEO of Neue Materialien Bayreuth GmbH, a Bavarian state R&D-institution in the field of materials and processes for polymers, composites and metals.

The research group of Volker Altstädt is dedicated to scientific and industrial-oriented research in the area of polymeric materials, establishing a connection between natural sciences and engineering technology. Emphasis is placed on the interdisciplinary cooperation among scientists, bringing together the disciplines of chemistry, physics, chemical engineering and mechanical engineering. Research activities of Volker Altstädt's group focus on polymer foams, polymer composites and nanocomposites, polymer blends and compatibilization, resin systems and flame protection, special injection molding techniques, with the primary goal of determining the structure–properties relationships and tailoring polymeric materials for specific requirements.

Preface to "Polymer Blends and Compatibilization"

The market is continuously looking for substitutes for expensive polymers or tailor made polymers for specific applications. Therefore, polymer blends are gaining more interest since they possess a great potential to fulfill these needs. Blending not only results in better final properties, but can also improve the processing behavior and reduce costs. In the field of polymer blends, there are numerous parameters that influence the morphology, e.g., viscosity ratio, blend composition, shear conditions, and blend ratio. There is still a great deal of potential to scientifically exploit the possibilities of blend technology, which is necessary to obtain a foundation based on science, engineering, technology, and applications in order to make it possible to tailor polymer blends as desired.

However, combining two or more different polymers to receive favorable properties by blending often results in immiscible polymer blends. This immiscibility goes hand-in-hand with phase separation leading to weak mechanical properties. The high interfacial tension causing this can be reduced by compatibilization of polymer blends. There are different methods to achieve this, such as adding block and graft copolymers, reactive polymers to form block and graft copolymers, nanoparticles or organic molecules. Using suitable compatibilizers, not only is the interfacial adhesion between matrix and its blends reduced, but also the dispersion of the dispersed phase is improved, the adhesion between the phases is enhanced and the morphology is stabilized. This can lead to improved mechanical and morphological properties.

Designing new polymer blends or improving the properties of immiscible polymer blends by compatibilization is very challenging, but an excellent way to exploit the full potential of polymers for applications and their varied needs. This Special Issue covers polymer blends from PC/ABS- over PLA/PBAT/PF- to PA6/PP-blends and showcases new aspects of compatibilization such as pre-crosslinking of carboxyl-terminated butadiene acrylonitrile liquid rubber/epoxy blends or the development of styrene-grafted polyurethane by radiation-based techniques.

Volker Altstädt
Guest Editor

![materials logo] MDPI

Article

Influence of Amphibian Antimicrobial Peptides and Short Lipopeptides on Bacterial Biofilms Formed on Contact Lenses

Magdalena Maciejewska [1,2,†], **Marta Bauer** [1,†], **Damian Neubauer** [1], **Wojciech Kamysz** [1] and **Malgorzata Dawgul** [1,*]

[1] Department of Inorganic Chemistry, Faculty of Pharmacy, Medical University of Gdansk, Gdansk 80-416, Poland; maciejewska.kj@gmail.com (M.M.); bauerm@gumed.edu.pl (M.B.); dneu@gumed.edu.pl (D.N.); kamysz@gumed.edu.pl (W.K.)
[2] Pharmaceutical Laboratory, Avena SJ, Osielsko 86-031, Poland
* Correspondence: mdawgul@gumed.edu.pl; Tel.: +48-58-349-1488 or +48-691-930-090
† These authors contributed equally to this work.

Academic Editor: Mauro Pollini
Received: 25 July 2016; Accepted: 20 October 2016; Published: 26 October 2016

Abstract: The widespread use of contact lenses is associated with several complications, including ocular biofilm-related infections. They are very difficult to manage with standard antimicrobial therapies, because bacterial growth in a biofilm is associated with an increased antibiotic resistance. The principal aim of this study was to evaluate the efficacy of antimicrobial peptides (AMPs) in eradication of bacterial biofilms formed on commercially available contact lenses. AMPs were synthesized according to Fmoc/tBu chemistry using the solid-phase method. Minimum inhibitory concentration (MIC) and minimum biofilm eradication concentration (MBEC) of the compounds were determined. Anti-biofilm activity of the antimicrobial peptides determined at different temperatures (25 °C and 37 °C) were compared with the effectiveness of commercially available contact lens solutions. All of the tested compounds exhibited stronger anti-biofilm properties as compared to those of the tested lens solutions. The strongest activity of AMPs was noticed against Gram-positive strains at a temperature of 25 °C. Conclusions: The results of our experiments encourage us toward further studies on AMPs and their potential application in the prophylaxis of contact lens-related eye infections.

Keywords: antimicrobial peptides; lipopeptides; biofilm; ocular infections; contact lenses

1. Introduction

Contact lenses (CL) are one of the most popular biomaterials and represent a suitable surface for bacterial colonization in the eye, leading to ocular infection. Approximately 140 million people worldwide wear CL for vision correction [1]. The widespread use of CL is associated with a risk of complications, including CL-related bacterial infections [2]. The ability of different species of bacteria to adhere to the surface of biomaterials resulting in biofilm formation plays an important role in pathogenesis of CL-associated infections. Microbial cells inhabiting a three dimensional structure of biofilm have been reported to be up to 1000 times more resistant to antibiotics than their planktonic forms [3].

The surface of the eye is constantly exposed to numerous environmental factors, including a frequent adhesion of microorganisms. Majority of the environmentally-introduced microorganisms are constantly removed from the surface of the eye through the natural host defense mechanisms, such as tears, corneal nerves, epithelial cells, keratocytes, interferons, and innate immunity cells. The primary line of defense against ocular infection is provided by mechanical barriers, eyelids, and eyelashes. The blinking action of the eyelids removes most foreign particles and microorganisms from the eye and renews the tear film. Tears are one of the crucial elements of the eye's defense system. They keep the eye lubricated and free from foreign bodies by flushing action. The tears' film also provides an effective antimicrobial system capable of reducing the number of bacteria in vivo.

The tear fluid is composed of water, electrolytes, lipids, mucins and proteins (immunoglobulins, lysozyme, lactoferrin, lipocalin, and lacritin) [4,5]. A combination of lysozyme and lactoferrin has been found to be highly effective in eradicating staphylococci [6].

The most common ocular infections associated with CL are bacterial conjunctivitis, contact lens acute red eye (CLARE), contact lens peripheral ulcer (CLPU), infiltrative keratitis (IK), and microbial keratitis (MK). *Staphylococcus aureus*, *S. epidermidis*, *Streptococcus pneumonia*, *S. pyogenes*, *Haemophilus influenzae*, *Enterococcus* spp. *Moraxella* spp. *Escherichia coli*, *Serratia marcescens*, *Pseudomonas aeruginosa*, and *Proteus mirabilis* are the most common etiological factors associated with CL-related infections [1,7,8].

Handling CL, not complying with the hygienic procedures, replacement frequency and a poor quality of material, inadequate care solutions, and CL case compliance are major factors of CL contamination. There is also a strong correlation between CL extended wear and corneal infection and inflammation [9]. Extended wear of CL promotes enhanced adhesion of bacteria and consequently the development of corneal infection [10].

The current standard treatment of bacterial eye infections relies on conventional antibiotic therapy. However, the widespread use of broad-spectrum systemic antibiotics in ocular infection therapies resulted in an increase in resistance among the bacteria and induction of antibiotic-resistant strains. A significant number of *S. aureus* and coagulase-negative staphylococci, as well as *P. aeruginosa* associated with ocular infection, have been strongly antibiotic-resistant [11,12]. The standard therapy is of limited effectiveness in abatement of biofilm infection.

Antimicrobial peptides (AMPs) have been found to exhibit activity against a wide range of bacteria responsible for ocular infections. These naturally occurring antimicrobial agents have been examined as therapeutic antibiotics. As a part of the innate immune system, AMPs can be applied to prevent the formation of biofilm as well as to eradicate mature structures. The mechanism of action of AMPs is based on the interaction with the cell membrane and carries a low risk of microbial resistance. AMP may be an alternative to conventional antibiotic treatment in ocular infection.

The most significant naturally occurring AMPs in the eye are human cathelicidin LL-37, as well as α- and β-defensins [4,5]. These compounds are secreted into tears by corneal and conjunctival epithelial cells and demonstrate a broad spectrum of antimicrobial activity. In response to inflammatory agents, neutrophils infiltrating the ocular surface supply tear fluid with additional LL-37 and α-defensins. *In vitro* studies confirmed antimicrobial activity of human β-defensins against methicillin-resistant *S. aureus* and *P. aeruginosa* in ocular infection [13]. The LL-37 is also produced by mast cells. The peptide significantly prevents from *S. epidermidis* adhesion and biofilm formation on the biomedical materials [5,13]. *In vitro* studies have shown that LL-37 is also effective in *P. aeruginosa* biofilm eradication [14]. Another AMP of tear fluid is psoriasin (S100A7). It is produced by cornea, conjunctiva, and lacrimal glands and has been demonstrated to be highly potent against *E. coli* [15].

Currently, several innovative strategies to optimize AMPs anti-biofilm activity are under evaluation with the aim to using them as a new class of antibiotics in the treatment of ocular infection. Direct topical AMPs application, production of AMP's derivatives and peptidomimetics, construction of specifically-targeted AMPs (STAMP), or immobilization of AMPs directly onto the surface of biomaterials are some of the examples [16,17]. Several in vitro and in vivo studies have been performed to investigate a group of AMPs as potential anti-biofilm agents for ocular infection therapies [17–21].

The principal aim of this work was to evaluate the group of antimicrobial peptides with regard to their potential use for eradication of bacterial biofilms formed on CL.

2. Results

2.1. Minimum Inhibitory Concentration of the AMPs

The AMPs under study exhibited antimicrobial activity towards all of the tested bacteria, but their efficacy varied depending on the strain (Table 1). Pexiganan turned out to be the most effective peptide against planktonic bacteria. Bacterial growth was inhibited after exposure to pexiganan at concentrations of 1–8 µg/mL. Short synthetic lipopeptides also effectively inhibited the growth of all tested strains (4–32 µg/mL). The remaining peptides showed a comparable activity towards Gram-positive strains (8–32 µg/mL), while application of significantly higher concentrations of the compounds was required to inhibit the growth of Gram-negative strains (128–512 µg/mL).

Table 1. Minimum inhibitory concentration of the AMPs in µg/mL (µM).

Peptide	S. aureus	S. epidermidis	S. pyogenes	S. pneumoniae	E. coli	P. aeruginosa
Pexiganan	8 (2.2)	2 (0.6)	4 (1.1)	≤1 (≤0.3)	4 (1.1)	4 (1.1)
Citropin 1.1	32 (17.4)	8 (4.3)	16 (8.7)	16 (8.7)	128 (69.5)	256 (138.9)
Temporin A	8 (4.9)	4 (2.5)	4 (2.5)	8 (4.9)	256 (157.5)	512 (315.1)
Palm-KK-NH$_2$	8 (10.8)	4 (5.4)	16 (21.7)	8 (10.8)	8 (10.8)	16 (21.7)
Palm-RR-NH$_2$	8 (10.1)	4 (5.0)	16 (20.1)	4 (5.0)	16 (20.1)	32 (40.3)

2.2. Anti-Biofilm Activity of Antimicrobial Peptides at Different Temperatures

The results of the tests with CL differed from those obtained for bacteria living in a planktonic form (Tables 2 and 3). MBEC tests have shown that citropin 1.1, the weakest antimicrobial agent against planktonic cells, was the most potent anti-biofilm agent. It was the only compound that eliminated the living bacteria from the surface of CL at both temperatures. The MBEC values ranged from 16–64 µg/mL for Gram-positive strains. Biofilms formed by *E. coli* were eradicated after exposure to citropin 1.1 at concentrations of 128–256 µg/mL. *P. aeruginosa* was eliminated from the lens surface after application of citropin 1.1 at concentration of 512 µg/mL. Additionally, temporin A eliminated the bacteria at that concentration. The remaining peptides were inactive towards *P. aeruginosa* cultured on CL. Promising results were obtained with lipopeptide Pal-KK-NH$_2$. The compound eradicated biofilms formed by Gram-positive bacteria at concentrations of 32–64 µg/mL below 25 °C and at 32–512 µg/mL at 37 °C. *E. coli* was eliminated by exposure to the compound at concentration of 256 µg/mL at both temperatures. The remaining compounds exhibited a moderate anti-biofilm activity. Surprisingly enough, pexiganan, the most potent antimicrobial agent in the MIC test, showed a rather feeble anti-biofilm activity. With certain strains and compounds the activity varied depending on the temperature. In many cases the MBEC values obtained for AMPs were 2–4 times lower at 25 °C than at 37 °C.

Table 2. Minimum biofilm eradication concentration (reduction of resazurin ≤10% ± 0.5%) at 37 °C in µg/mL (µM).

Peptide	S. aureus	S. epidermidis	S. pyogenes	S. pneumoniae	E. coli	P. aeruginosa
Pexiganan	>512 (>141.5)	128 (35.4)	256 (70.8)	64 (17.7)	>512 (>141.5)	>512 (>141.5)
Citropin 1.1	64 (34.7)	16 (8.7)	32 (17.4)	32 (17.4)	256 (138.9)	512 (277.8)
Temporin A	512 (315.1)	128 (78.8)	128 (78.8)	128 (78.8)	512 (315.1)	512 (315.1)
Palm-KK-NH$_2$	64 (86.6)	32 (43.3)	512 (693.0)	64 (86.6)	256 (346.5)	>512 (>693.0)
Palm-RR-NH$_2$	256 (322.1)	32 (40.3)	>512 (>644.1)	32 (40.3)	512 (644.1)	>512 (>644.1)

Table 3. Minimum biofilm eradication concentration (reduction of resazurin ≤10% ± 0.5%) at 25 °C (µg/mL (µM)).

Peptide	S. aureus	S. epidermidis	S. pyogenes	S. pneumoniae	E. coli	P. aeruginosa
Pexiganan	256 (70.8)	16 (4.4)	128 (35.4)	64 (17.7)	512 (141.5)	>512 (>141.5)
Citropin 1.1	32 (17.4)	16 (8.7)	32 (17.4)	64 (34.7)	128 (69.5)	512 (277.8)
Temporin A	128 (78.8)	64 (39.4)	256 (157.5)	128 (78.8)	128 (78.8)	512 (315.1)
Palm-KK-NH$_2$	32 (43.3)	32 (43.3)	32 (43.3)	64 (86.6)	256 (346.5)	>512 (>693.0)
Palm-RR-NH$_2$	128 (161.0)	64 (80.5)	128 (161.0)	64 (80.5)	256 (322.1)	>512 (>644.1)

2.3. Efficacy of the Contact Lens Solutions against Biofilms

The commercial contact lens solutions were more effective after a single application at 37 °C (Table 4). This was observed for three out of four tested solutions against *S. pyogenes* and *E. coli*, and for one solution in case of *P. aeruginosa*. The CL solutions were active at both temperatures against biofilms formed on the lens surface by *S. aureus*, *S. epidermidis*, and *S. pneumoniae*. They were ineffective against *E. coli* and *S. pyogenes* cultured on lenses at 25 °C. The most difficult to eliminate from the surface of lenses was *P. aeruginosa*. It could not be eradicated after exposure to all solutions at 25 °C and to three out of four solutions at 37 °C.

Table 4. Efficacy of the contact lens solutions against biofilm at different temperatures (37 °C/ 25 °C).

Lens Solution	S. aureus	S. epidermidis	S. pyogenes	S. pneumoniae	E. coli	P. aeruginosa
A	+/+	+/+	−/−	+/+	+/−	−/−
B	+/+	+/+	+/−	+/+	+/+	+/−
C	+/+	+/+	+/−	+/+	+/−	−/−
D	+/+	+/+	+/−	+/+	+/−	−/−

"+" active; reduction of resazurin ≤10% ± 0.5%, "−" inactive; reduction of resazurin ≥10% ± 0.5%.

3. Discussion

Biomaterials associated infections constitute a major barrier to the long-term use of medical devices and remain a serious therapeutic problem [17,22,23]. Bacterial growth in a form of biofilm is associated with an increased antibiotic resistance as compared to the growth under planktonic conditions [24].

CL represent an ideal surface for bacterial colonization in the eye. CL-related infections are relatively rare, but can pose severe vision-threatening complications. Standard treatment for most bacterial ocular infection is primarily empiric with broad-spectrum antibiotics. Unfortunately, the inadequate use of antibiotics leads to the development of resistant strains [25–27].

An alternative approach to the therapeutic problem of biofilm-associated infection would be the development of a new generation of AMP-based drugs [28,29]. In this study we confirmed antibacterial activity of the AMPs against common strains responsible for ocular infections.

In the MIC test the most potent antimicrobial agent was pexiganan, an analogue of the naturally occurring magainin 2. Its strong antimicrobial activity has been reported by numerous studies [30–32]. The results clearly show the high effectiveness and broad-spectrum activity of pexiganan [33]. Potential therapeutic applications of the peptide and further development of its analogues are under way [34–36]. Although the compound has shown a broad spectrum of antimicrobial activity against planktonic cells of various bacterial strains, pexiganan turned out to be a weak anti-biofilm agent against structures formed on CL. This supports the finding that bacteria organized in a biofilm can exhibit a reduced susceptibility to AMPs. In our previous study reference bacterial strains susceptible to conventional antibiotics in their planktonic form were highly resistant once cultured as biofilms on polystyrene surfaces [32,37].

Both tested short lipopeptides also exhibited a high antimicrobial activity against all of the tested strains in the MIC test. Previous studies on Palm-KK-NH$_2$ showed its strong activity against clinical strains of Gram-positive cocci including those resistant to antibiotics [38]. Pal-KK-NH$_2$ demonstrated its ability to prevent a vascular graft biofilm formation in the rat model of staphylococcal infections. Moreover, efficiency of the lipopeptide was enhanced upon combination with standard drugs (e.g., vancomycin) [39]. Pal-KK-NH$_2$ was also effective against a majority of strains in the biofilm form, while Pal-RR-NH$_2$ showed a slightly lower activity. In our previous study the compounds were effective against biofilms formed by clinical isolates of *S. aureus* cultured on polystyrene [40].

Two amphibian peptides, citropin 1.1, naturally produced by *Litoria citropa* [41], and temporin A, primarily isolated from *Rana temporaria* [42], were highly effective against planktonic cultures of Gram-positive bacteria [43–46]. In the previous studies, temporin A also exhibited a strong activity against Gram-positive bacteria, including vancomycin-resistant strains [47]. Effectiveness of the compound was confirmed in animal models of infection caused by *S. epidermidis* [48] and *S. aureus* [49]. Research conducted by Simonetti et al. confirmed that temporin A was effective in MRSA-infected chronic wounds in mice [50]. Citropin 1.1 was also extensively studied. Most recently, it was found to be a potent anti-biofilm agent against *S. aureus* strains collected from patients with atopic dermatitis [51]. Moreover, the previous studies confirmed that the peptide demonstrates a synergy with rifampicin and minocycline against *S. aureus* biofilm [52]. In another study, citropin 1.1 showed a high efficiency in prevention of catheter-related infections caused by *S. aureus* [52]. Our results have shown that both compounds eliminated *P. aeruginosa* from CL, in contrary to the remaining peptides and applied CL solutions. Citropin 1.1 was the most effective compound in the study against biofilms formed by all of the tested strains.

The obtained results indicate that AMPs possess a potential for the development as a therapeutic and prophylactic agents against biofilm-related infections, especially those associated with Gram-positive pathogens.

In recent years a variety of peptides were investigated for their potential in the prevention of eye infections. For instance, Nos–Barbera et al. demonstrated positive effects of a cecropin-melittin hybrid peptide topically applied in a pseudomonas keratitis model in rabbits [18]. Willcox et al. developed antimicrobial cationic peptide coatings by covalent immobilization of melimine onto CL. Melimine is a hybrid of the active regions of two antimicrobial peptides: protamine (from salmon sperm) and melittin (from bee venom). Covalently-attached melimine significantly reduced clinical manifestations of CLARE in the *P. aeruginosa* guinea pig model [17]. *In vivo* melimine coatings demonstrated a significant reduction of CLPU incidence in the rabbit model [19]. The broad-spectrum antimicrobial activity of melimine was also confirmed in a human clinical trial, where melimine-coated lenses turned out to be safe for use during wear and retained a high antimicrobial activity after wear [20]. Other studies have shown that the melimine coating was heat-stable, non-toxic to mammalian cells in vitro, and did not alter the physical dimensions of CL [17,19,21].

In this study, we investigated peptides according to their potential application as CL solutions additives. All peptides eliminated Gram-positive bacteria from the surface of lenses, while only temporin A and citropin 1.1 showed some activity against Gram-negative bacteria. All of the tested CL solutions were ineffective against *S. pyogenes* and *P. aeruginosa* when cultured once at 25 °C. Three out of four applied solutions were also inactive towards *E. coli*. Similarly to previous studies, we have now confirmed that the efficacy of CL disinfecting solutions are dependent on the type of the bacterial strain [53–55]. The results of our study indicate that *P. aeruginosa* was the most resistant strain to the antimicrobial action of all of the tested fluids, a finding consistent with those of other studies [56,57].

Our research indicates that the susceptibility to antimicrobials of bacteria cultured on the lens surfaces is dependent on the temperature. For some strains and the tested AMPs, the anti-biofilm activity was stronger at 25 °C, while the tested lens solutions exhibited stronger activity at 37 °C. As room temperature is commonly recommended for the storage of CL, the use of commercial CL solutions appears to be rather inadequate for the prevention of ocular infections in CL users.

A harmony of proper hygiene and care habits of CL can effectively prevent contamination with pathogenic microorganisms [58]. Handling of CL, insufficient hand washing practice, and failure of some preservative systems are major sources of contamination implicated in the development of CL-associated infections [59–61]. According to the report of Turner et al., 44% of CL users do not wash their hands before handling, significantly increasing the risk of introducing pathogenic microorganisms into the eye [62]. The CL storage case and solutions may also be potential reservoirs for microorganisms responsible for ocular infections [5]. For example, studies exploring bacterial bioburden in a CL care system confirmed that storage case contamination is much more common than associated CL contamination [63–65]. The presence of pathogens in a storage case and in CL solutions significantly increase the risk of their transfer to the lens surface and, ultimately, onto the surface of the eye.

Our data reveal that AMPs might be promising antibacterial additives to CL solutions. However there are certain limitations and issues which need to be considered. The long-time stability of compounds in a water solution needs to be examined, but first of all the evaluation of toxicity towards human eye cells needs to be carried out. In the previous study tested amphibian peptides turned out to be safe towards HaCaT cells at their microbiologically-active concentrations. In the same assay lipopeptide Pal-KK-NH$_2$ exhibited high toxicity towards human keratinocytes [66]. Considering the potential toxicity and the results of the present study, the most interesting candidates for further examination are citropin 1.1 and temporin A. Encouraging results for other AMPs have been previously reported. Mannis et al. evaluated a synthetic cecropin analogue (Shiva-11) as an antibacterial agent in CL solutions. They demonstrated that it was effective against *P. aeruginosa*, *S. epidermidis* and *S. aureus* in a buffered saline containing a CL [67]. In an extensive study of the Shiva-11, Gunshefski et al. revealed that the peptide has a wide range of antimicrobial activity in vitro against human clinical ocular pathogens [68].

Sousa et al. examined the effectiveness of a cecropin analogue (D$_5$C) against *P. aeruginosa* and compared it with the antimicrobial effect of commercial CL solutions. The investigators demonstrated that D$_5$C substantially enhanced antimicrobial activity of the disinfecting solutions [69].

The results obtained in this study are consistent with those previously reported. AMPs show a broad-range antimicrobial activity and are potent anti-biofilm agents. Based on our results and the literature data, we came to the conclusion that AMPs show a considerable potential for the development of therapeutic and prophylactic antimicrobials against CL related infections. Further research is needed to establish the safety of ophthalmic application of the tested AMPs, as well as to optimize the composition of CL solutions, to ensure stability of the applied AMPs. Effective products, in combination with education on hygiene recommendations, will contribute to minimizing the risk of eye infections.

4. Materials and Methods

4.1. Bacterial Strains and Culture Conditions

All selected strains were obtained from the Polish Collection of Microorganisms (Polish Academy of Science, Wroclaw, Poland). Four Gram-positive and three Gram-negative strains linked with CL-related infections lenses were tested (Table 5). The microorganisms were cultured in a Mueller Hinton Broth II (MHB, Biocorp, Warsaw, Poland) overnight, at 37 °C.

Table 5. Reference bacterial strains.

Bacterial Group	Species	Number
Gram-positive	*Staphylococcus aureus*	ATCC 25923
	Staphylococcus epidermidis	ATCC 14990
	Streptococcus pneumoniae	ATCC 49619
	Streptococcus pyogenes	PCM 465
Gram-negative	*Escherichia coli*	ATCC 25922
	Pseudomonas aeruginosa	ATCC 9027

4.2. Antimicrobial Peptides

All of the peptides were synthesized in the Department of Inorganic Chemistry (Medical University of Gdansk, Gdansk, Poland). These were:

Pexiganan: GIGKFLKKAKKFGKAFVKILKK-NH$_2$
Citropin 1.1: GLFDVIKKVASVIGGL-NH$_2$
Temporin A: FLPLIGRVLSGIL-NH$_2$
Lipopeptides: Palm-KK-NH$_2$ and Palm-RR-NH$_2$ (Palm—hexadecanoic acid residue)

4.3. Contact Lenses and Contact Lens Solutions

The CL and CL solutions used are commercially available. One-day CL (1-Day Acuvue Moist) containing Etafilcon A were obtained from Johnson and Johnson Vision Care (Jacksonville, FL, USA). Four popular CL solutions were also tested (Table 6).

Table 6. Contact lens solutions and their contents.

Lens Solution	Components
A	Polyhexanide 0.0001%, Hydrolock (dexpanthenol and sorbitol), sodium phosphate, tromethamine, Poloxamer 407, disodium edetate.
B	Citrate, Tetronic 1304, aminomethylpropanol, sodium chloride, boric acid, sorbitol, disodium edetate, Polyquad (Polyquaternium) 0.001%, Aldox (myristamidopropyl dimethylamine) 0.0005%.
C	Hydroxyphosphate Alkyl, Boric Acid 0.03%, Disodium Edetate, Poloxamine 1%, Sodium Borate, Sodium Chloride, Polyaminopropyl Biguanide 0.0001%.
D	Boric Acid, disodium edetate, sodium borate, sodium chloride, DYMED (polyaminopropyl biguanide) 0.0001%, HYDRANATE (hydroxyalkylphosphonate) 0.03%, Poloxamine 1%.

4.4. Peptide Synthesis

All of the peptides and lipopeptides were synthesized manually on polystyrene resin (Orpegen, Heidelberg, Germany) modified by a Rink Amide linker [70]. 9-Fluorenylmethoxycarbonyl (Fmoc) groups were used for protection of the α-amino groups of amino acids. Deprotection of the groups was carried out for 20 min using a 20% piperidine (Merck, Darmstadt, Germany) solution in *N,N*-dimethylformamide (DMF) (Honeywell, Seelze, Germany). Then the resin was washed in DMF and dichloromethane (DCM) (Chempur, Piekary Slaskie, Poland). To check the completion of the deprotection and acylation process, a chloranil test was performed. All amino acids (Orpegen, Heidelberg, Germany) were coupled using a DMF/DCM (1:1 *v/v*) mixture in the presence of coupling agents, 1-hydroxybenzotriazole (HOBt) (Orpegen, Heidelberg, Germany) and *N,N*-diisopropylocarbodiimide (DIC) (Merck, Darmstadt, Germany). The progress of each acylation was controlled by a chloranil test. The peptides were cleaved from the resin with a mixture of trifluoroacetic acid (TFA) (Merck, Darmstadt, Germany), water, triisopropylsilane (TIS) (Merck, Darmstadt, Germany), and phenol (Sigma Aldrich, Steinheim, Germany) as scavengers in the ratio 92.5:2.5:2.5:2.5 *v/v*. Once the sequences were obtained, the peptides were precipitated with cold diethyl ether (Chempur, Piekary Slaskie, Poland) and lyophilized. The crude products were purified and analyzed by reversed-phase high-performance liquid chromatography (RP-HPLC) in an acetonitrile-water (Sigma Aldrich, Steinheim, Germany) gradient containing 0.1% TFA. The identity of the peptides was confirmed by matrix-assisted laser desorption/ionization time of flight mass spectrometry (MALDI-TOF) and the counter-ion was determined by ion chromatography [71,72]. MS and HPLC analysis are included in Supplementary Materials.

4.5. Antimicrobial Activity Test Protocol

Peptide susceptibility tests on the reference bacterial strains were performed with the following procedures recommended by the Clinical and Laboratory Standard Institute (CLSI). The activity of five peptides was tested on the Mueller Hinton Broth II (Biocorp, Warsaw, Poland) using broth microdilution method on polystyrene 96-well plates (Kartell, Noviglio, Italy). Bacterial inoculum of ca. 5×10^5 CFU/mL was added to each well and exposed to solutions of peptides at increasing concentrations (range 1–512 µg/mL). The plates were incubated overnight at 37 °C. MIC was assumed as the lowest concentration at which bacterial growth was no longer visible. All experiments were performed in triplicate.

4.6. Biofilm Assay

Biofilms of strains related with ocular infections were cultured on CL placed in polystyrene 24-well plates (Orange Scientific, Braine-l'Alleud, Belgium). Biomaterials were incubated at 37 °C in bacterial suspension in the Mueller Hinton Broth II (Biocorp, Warsaw, Poland) at initial inoculums of ca. 5×10^5 CFU/mL. After 24 h of incubation all of the CL were rinsed three times with a sterile phosphate buffer (AppliChem, Darmstadt, Germany). The lenses were then transferred into new wells with fresh MHB II. One-day biofilms were exposed to a pre-determined range of peptides concentration. Plates with antimicrobial agents were incubated again for 24 h at 37 °C (optimal for bacterial growth) or 25 °C (room temperature-standard storage temperature for CL). Anti-biofilm activity of selected peptides was visualized using a cell-viability reagent, Resazurin (Sigma Aldrich, St. Louis, MO, USA). Upon contact with living cells the dye is metabolized by bacterial dehydrogenases resulting in the reduction of the blue resazurin to a pink resorufin. Positive controls contained CL immersed in bacterial suspensions without antimicrobials, while CL in a sterile culture medium served as negative controls. The minimum biofilm eradication concentration (MBEC) was read after a one-hour incubation with Resazurin (final concentration per sample = 0.005%) MBEC was determined as the lowest concentration at which the reduction of resazurin was lower or equal (10% ± 0.5%) as compared to positive (100%) and negative (0%) controls. All experiments were performed in triplicate.

4.7. Antimicrobial Activity of the Contact Lens Solutions

Effectiveness of the CL solutions against biofilms formed on CL was tested using the previously-described procedure, with the difference that in the place of peptides, different undiluted CL solutions were applied. Tested solutions were considered as active when they demonstrated the ability to reduce the number of living bacteria on the surface of CL by at least 90% (reduction of resazurin to lower or equal to (10% $\pm$ 0.5%) as compared to positive (100%) and negative (0%) controls). All experiments were performed in triplicate.

Supplementary Materials: MS and HPLC analysis are available online at www.mdpi.com/1996-1944/9/11/873/s1.

Acknowledgments: This research was supported by a grant from the Polish National Science Centre (No. 2011/03/B/NZ7/00548) and a grant from the Ministry of Science and Higher Education of the Republic of Poland within the framework of the quality-promoting subsidy, under the Leading National Research Centre (KNOW) program for the years 2012–2017.

Author Contributions: M.M. performed the microbiological experiments and wrote the paper; M.B. performed the microbiological experiments and helped with preparation of the manuscript for publication; D.N. performed synthesis and purification of tested peptides; W.K. contributed materials and analysis tools and supervised the peptide synthesis and microbiological assays; M.D. designed the experiments, analyzed the data and supervised preparation of the manuscript for publication.

Conflicts of Interest: The authors declare no conflict of interest.

References

1. Bennet, E.S. GP Annual Report. Contact Lens Spectrum. Available online: http://www.clspectrum.com (accessed on 23 June 2012).
2. Szczotka-Flynn, L.B.; Pearlman, E.; Ghannoum, M. Microbial contamination of contact lenses, lens care solutions, and their accessories: A literature review. *Eye Contact Lens* **2010**, *36*, 116–129. [CrossRef] [PubMed]
3. Costerton, J.W.; Stewart, P.S.; Greenberg, E.P. Bacterial biofilms: A common cause of persistent infections. *Science* **1999**, *284*, 1318–1322. [CrossRef] [PubMed]
4. McDermott, A.M. Antimicrobial compounds in tears. *Exp. Eye Res.* **2013**, *117*, 53–61. [CrossRef] [PubMed]
5. McDermott, A.M. The role of antimicrobial peptides at the ocular surface. *Ophthalmic Res.* **2009**, *41*, 60–75. [CrossRef] [PubMed]
6. Leitch, E.C.; Willcox, M.D. Elucidation of the antistaphylococcal action of lactoferrin and lysozyme. *J. Med. Microbiol.* **1999**, *48*, 867–871. [CrossRef] [PubMed]
7. Willcox, M.D.; Holden, B.A. Contact lens related corneal infections. *Biosci. Rep.* **2001**, *21*, 445–461. [CrossRef] [PubMed]
8. Wu, P.; Stapleton, F.; Willcox, M.D. The causes of and cures for contact lens-induced peripheral ulcer. *Eye Contact Lens* **2003**, *29*, S63–S66, S83–S84, S192–S194. [CrossRef] [PubMed]
9. Weissman, B.; Mondino, B.J. Why daily wear is still better than extended wear. *Eye Contact Lens* **2003**, *29*, S145–S146, S166, S192–S194. [CrossRef] [PubMed]
10. Keay, L.; Stapleton, F.; Schein, O. Epidemiology of contact lens-related inflammation and microbial keratitis: A 20-year perspective. *Eye Contact Lens* **2007**, *33*, 346–353. [CrossRef] [PubMed]
11. Haas, W.; Pillar, C.M.; Torres, M.; Morris, T.W.; Sahm, D.F. Monitoring antibiotic resistance in ocular microorganisms: Results from the antibiotic resistance monitoring in ocular micRorganisms (ARMOR) 2009 surveillance study. *Am. J. Ophthalmol.* **2011**, *152*, 567–574. [CrossRef] [PubMed]
12. Willcox, M.D. Review of resistance of ocular isolates of *Pseudomonas aeruginosa* and staphylococci from keratitis to ciprofloxacin, gentamicin and cephalosporins. *Clin. Exp. Optom.* **2011**, *94*, 161–168. [CrossRef] [PubMed]
13. Hell, E.; Giske, C.G.; Nelson, A.; Römling, U.; Marchini, G. Human cathelicidin peptide LL37 inhibits both attachment capability and biofilm formation of *Staphylococcus epidermidis*. *Lett. Appl. Microbiol.* **2010**, *50*, 211–215. [CrossRef] [PubMed]
14. Kolar, S.S.; McDermott, A.M. Role of host-defence peptides in eye diseases. *Cell Mol. Life Sci.* **2011**, *68*, 2201–2213. [CrossRef] [PubMed]

15. Garreis, F.; Gottschalt, M.; Schlorf, T.; Gläser, R.; Harder, J.; Worlitzsch, D.; Paulsen, F.P. Expression and regulation of antimicrobial peptide psoriasin (S100A7) at the ocular surface and in the lacrimal apparatus. *Investig. Ophthalmol. Vis. Sci.* **2011**, *52*, 4914–4922. [CrossRef] [PubMed]

16. Batoni, G.; Maisetta, G.; Brancatisano, F.L.; Esin, S.; Campa, M. Use of antimicrobial peptides against microbial biofilms: Advantages and limits. *Curr. Med. Chem.* **2011**, *18*, 256–279. [CrossRef] [PubMed]

17. Willcox, M.D.; Hume, E.B.; Aliwarga, Y.; Kumar, N.; Cole, N. A novel cationic-peptide coating for the prevention of microbial colonization on contact lenses. *J. Appl. Microbiol.* **2008**, *105*, 1817–1825. [CrossRef] [PubMed]

18. Nos-Barbera, S.; Portoles, M.; Morilla, A.; Ubach, J.; Andreu, D.; Paterson, C.A. Effect of hybrid peptides of cecropin A and melittin in an experimental model of bacterial keratitis. *Cornea* **1997**, *16*, 101–106. [CrossRef] [PubMed]

19. Cole, N.; Hume, E.B.; Vijay, A.K.; Sankaridurg, P.; Kumar, N.; Willcox, M.D. In vivo performance of melimine as an antimicrobial coating for contact lenses in models of CLARE and CLPU. *Investig. Ophthalmol. Vis. Sci.* **2010**, *51*, 390–395. [CrossRef] [PubMed]

20. Dutta, D.; Ozkan, J.; Willcox, M.D. Biocompatibility of antimicrobial melimine lenses: Rabbit and human studies. *Optom. Vis. Sci.* **2014**, *91*, 570–581. [CrossRef] [PubMed]

21. Dutta, D.; Cole, N.; Kumar, N.; Willcox, M.D. Broad spectrum antimicrobial activity of melimine covalently bound to contact lenses. *Investig. Ophthalmol. Vis. Sci.* **2013**, *54*, 175–182. [CrossRef] [PubMed]

22. Von Eiff, C.; Jansen, B.; Kohnen, W.; Becker, K. Infections associated with medical devices: Pathogenesis, management and prophylaxis. *Drugs* **2005**, *65*, 179–214. [CrossRef] [PubMed]

23. Khardori, N.; Yassien, M. Biofilms in device-related infections. *J. Ind. Microbiol.* **1995**, *15*, 141–147. [CrossRef] [PubMed]

24. Mah, T.F.; O'Toole, G.A. Mechanisms of biofilm resistance to antimicrobial agents. *Trends Microbiol.* **2001**, *9*, 34–39. [CrossRef]

25. Willcox, M.D.; Harmis, N.C.; Williams, T.H. Bacterial interactions with contact lenses; effects of lens material, lens wear and microbial physiology. *Biomaterials* **2001**, *22*, 3235–3247. [CrossRef]

26. Dutta, D.; Cole, N.; Willcox, M. Factors influencing bacterial adhesion to contact lenses. *Mol. Vis.* **2012**, *18*, 14–21. [PubMed]

27. Bertino, J.S., Jr. Impact of antibiotic resistance in the management of ocular infections: The role of current and future antibiotics. *Clin. Ophthalmol.* **2009**, *3*, 507–521. [CrossRef] [PubMed]

28. Kazemzadeh-Narbat, M.; Lai, B.F.; Ding, C.; Kizhakkedathu, J.N.; Hancock, R.E.; Wang, R. Multilayered coating on titanium for controlled release of antimicrobial peptides for the prevention of implant-associated infections. *Biomaterials* **2013**, *34*, 5969–5977. [CrossRef] [PubMed]

29. McDermott, A.M. Defensins and other antimicrobial peptides at the ocular surface. *Ocul. Surf.* **2004**, *2*, 229–247. [CrossRef]

30. Flamm, R.K.; Rhomberg, P.R.; Farrell, D.J.; Jones, R.N. In vitro spectrum of pexiganan activity; bactericidal action and resistance selection tested against pathogens with elevated MIC values to topical agents. *Diagn. Microbiol. Infect. Dis.* **2016**, *86*, 66–69. [CrossRef] [PubMed]

31. Cirioni, O.; Simonetti, O.; Pierpaoli, E.; Barucca, A.; Ghiselli, R.; Orlando, F.; Pelloni, M.; Minardi, D.; Trombettoni, M.M.; Guerrieri, M.; et al. Enhanced Efficacy of Combinations of Pexiganan with Colistin Versus Acinetobacter Baumannii in Experimental Sepsis. *Shock* **2016**, *46*, 219–225. [CrossRef] [PubMed]

32. Dawgul, M.; Maciejewska, M.; Jaskiewicz, M.; Karafova, A.; Kamysz, W. Antimicrobial peptides as potential tool to fight bacterial biofilm. *Acta Pol. Pharm.* **2014**, *71*, 39–47. [PubMed]

33. Maloy, W.L.; Kari, U.P. Structure-activity studies on magainins and other host defense peptides. *Biopolymers* **1995**, *37*, 105–122. [CrossRef] [PubMed]

34. Radzishevsky, I.S.; Rotem, S.; Zaknoon, F.; Gaidukov, L.; Dagan, A.; Mor, A. Effects of acyl versus aminoacyl conjugation on the properties of antimicrobial peptides. *Antimicrob. Agents Chemother.* **2005**, *49*, 2412–2420. [CrossRef] [PubMed]

35. Thennarasu, S.; Lee, D.K.; Tan, A.; Prasad, K.U.; Ramamoorthy, A. Antimicrobial activity and membrane selective interactions of a synthetic lipopeptide MSI-843. *Biochim. Biophys. Acta* **2005**, *1711*, 49–58. [CrossRef] [PubMed]

36. Meng, H.; Kumar, K. Antimicrobial activity and protease stability of peptides containing fluorinated amino acids. *J. Am. Chem. Soc.* **2007**, *129*, 15615–15622. [CrossRef] [PubMed]

37. O'Toole, G.; Kaplan, H.B.; Kolter, R. Biofilm formation as microbial development. *Annu. Rev. Microbiol.* **2000**, *54*, 49–79. [CrossRef] [PubMed]

38. Kamysz, W.; Silvestri, C.; Cirioni, O.; Giacometti, A.; Licci, A.; Della, V.A.; Okroj, M.; Scalise, G. In vitro activities of the lipopeptides palmitoyl (Pal)-Lys-Lys-NH(2) and Pal-Lys-Lys alone and in combination with antimicrobial agents against multiresistant gram-positive cocci. *Antimicrob. Agents Chemother.* **2007**, *51*, 354–358. [CrossRef] [PubMed]

39. Cirioni, O.; Giacometti, A.; Ghiselli, R.; Kamysz, W.; Silvestri, C.; Orlando, F.; Mocchegiani, F.; Vittoria, A.D.; Kamysz, E.; Saba, V.; et al. The lipopeptides Pal-Lys-Lys-NH(2) and Pal-Lys-Lys soaking alone and in combination with intraperitoneal vancomycin prevent vascular graft biofilm in a subcutaneous rat pouch model of staphylococcal infection. *Peptides* **2007**, *28*, 1299–1303. [CrossRef] [PubMed]

40. Dawgul, M.; Baranska-Rybak, W.; Kamysz, E.; Karafova, A.; Nowicki, R.; Kamysz, W. Activity of short lipopeptides and conventional antimicrobials against planktonic cells and biofilms formed by clinical strains of *Staphylococcus aureus*. *Future Med. Chem.* **2012**, *4*, 1541–1551. [CrossRef] [PubMed]

41. Wegener, K.L.; Wabnitz, P.A.; Carver, J.A.; Bowie, J.H.; Chia, B.C.; Wallace, J.C.; Tyler, M.J. Host defence peptides from the skin glands of the Australian blue mountains tree-frog Litoria citropa. Solution structure of the antibacterial peptide citropin 1.1. *Eur. J. Biochem.* **1999**, *265*, 627–637. [CrossRef] [PubMed]

42. Simmaco, M.; Mignogna, G.; Canofeni, S.; Miele, R.; Mangoni, M.L.; Barra, D. Temporins, antimicrobial peptides from the European red frog Rana temporaria. *Eur. J. Biochem.* **1996**, *242*, 788–792. [CrossRef] [PubMed]

43. Wade, D.; Silberring, J.; Soliymani, R.; Heikkinen, S.; Kilpeläinen, I.; Lankinen, H.; Kuusela, P. Antibacterial activities of temporin A analogs. *FEBS Lett.* **2000**, *479*, 6–9. [CrossRef]

44. Baranska-Rybak, W.; Cirioni, O.; Dawgul, M.; Sokolowska-Wojdylo, M.; Naumiuk, L.; Szczerkowska-Dobosz, A.; Nowicki, R.; Roszkiewicz, J.; Kamysz, W. Activity of antimicrobial peptides and conventional antibiotics against superantigen positive *Staphylococcus aureus* isolated from the patients with neoplastic and inflammatory erythrodermia. *Chemother. Res. Pract.* **2011**, *2011*, 270932. [PubMed]

45. Kamysz, W. Przeciwbakteryjna aktywność pepetydów ze skóry płazów. *Ann. Acad. Med. Gedan.* **2005**, *35*, 29–34.

46. Kamysz, W.; Mickiewicz, B.; Rodziewicz-Motowidło, S.; Greber, K.; Okrój, M. Temporin A and its retro-analogues: Synthesis, conformational analysis and antimicrobial activities. *J. Pept. Sci.* **2006**, *12*, 533–537. [CrossRef] [PubMed]

47. Cirioni, O.; Giacometti, A.; Ghiselli, R.; Dell'Acqua, G.; Gov, Y.; Kamysz, W.; Lukasiak, J.; Mocchegiani, F.; Orlando, F.; D'Amato, G.; et al. Prophylactic efficacy of topical temporin A and RNAIII-inhibiting peptide in a subcutaneous rat Pouch model of graft infection attributable to staphylococci with intermediate resistance to glycopeptides. *Circulation* **2003**, *108*, 767–771. [CrossRef] [PubMed]

48. Giacometti, A.; Cirioni, O.; Ghiselli, R.; Orlando, F.; D'Amato, G.; Kamysz, W.; Mocchegiani, F.; Sisti, V.; Silvestri, C.; Łukasiak, J.; et al. Temporin A soaking in combination with intraperitoneal linezolid prevents vascular graft infection in a subcutaneous rat pouch model of infection with *Staphylococcus epidermidis* with intermediate resistance to glycopeptides. *Antimicrob. Agents Chemother.* **2004**, *48*, 3162–3164. [CrossRef] [PubMed]

49. Cirioni, O.; Giacometti, A.; Ghiselli, R.; Kamysz, W.; Orlando, F.; Mocchegiani, F.; Silvestri, C.; Licci, A.; Łukasiak, J.; Saba, V.; et al. Temporin A alone and in combination with imipenem reduces lethality in a mouse model of staphylococcal sepsis. *J. Infect. Dis.* **2005**, *192*, 1613–1620. [CrossRef] [PubMed]

50. Simonetti, O.; Cirioni, O.; Goteri, G.; Ghiselli, R.; Kamysz, W.; Kamysz, E.; Silvestri, C.; Orlando, F.; Barucca, C.; Scalise, A.; et al. Temporin A is effective in MRSA-infected wounds through bactericidal activity and acceleration of wound repair in a murine model. *Peptides* **2008**, *29*, 520–528. [CrossRef] [PubMed]

51. Dawgul, M.; Baranska-Rybak, W.; Piechowicz, L.; Bauer, M.; Neubauer, D.; Nowicki, R.; Kamysz, W. The antistaphylococcal activity of citropin 1.1 and temporin a against planktonic cells and biofilms formed by isolates from patients with atopic dermatitis: An assessment of their potential to induce microbial resistance compared to conventional antimicrobials. *Pharmaceuticals* **2016**, *9*, E30. [PubMed]

52. Cirioni, O.; Giacometti, A.; Ghiselli, R.; Kamysz, W.; Orlando, F.; Mocchegiani, F.; Silvestri, C.; Licci, A.; Chiodi, L.; Lukasiak, J.; et al. Citropin 1.1-treated central venous catheters improve the efficacy of hydrophobic antibiotics in the treatment of experimental staphylococcal catheter-related infection. *Peptides* **2006**, *27*, 1210–1216. [CrossRef] [PubMed]

53. Hume, E.B.; Flanagan, J.; Masoudi, S.; Zhu, H.; Cole, N.; Willcox, M.D. Soft contact lens disinfection solution efficacy: Clinical Fusarium isolates vs. ATCC 36031. *Optom. Vis. Sci.* **2009**, *86*, 415–419. [CrossRef] [PubMed]

54. Hume, E.B.; Zhu, H.; Cole, N.; Huynh, C.; Lam, S.; Willcox, M.D. Efficacy of contact lens multipurpose solutions against serratia marcescens. *Optom. Vis. Sci.* **2007**, *84*, 316–320. [CrossRef] [PubMed]

55. Contact Lens Spectrum. Testing Multi-Purpose Lens Care Solution against *Staphylococcus aureus*. Available online: http://www.clspectrum.com/articleViewer.aspx?articleID=13228 (accessed on 1 April 2007).

56. Mohammadinia, M.; Rahmani, S.; Eslami, G.; Ghassemi-Broumand, M.; Aghazadh Amiri, M.; Aghaie, G.; Tabatabaee, S.M.; Taheri, S.; Behgozin, A. Contact lens disinfecting solutions antibacterial efficacy: Comparison between clinical isolates and the standard ISO ATCC strains of *Pseudomonas aeruginosa* and *Staphylococcus aureus*. *Eye* **2012**, *26*, 327–330. [CrossRef] [PubMed]

57. Manuj, K.; Gunderson, C.; Troupe, J.; Huber, M.E. Efficacy of contact lens disinfecting solutions against *Staphylococcus aureus* and *Pseudomonas aeruginosa*. *Eye Contact Lens* **2006**, *32*, 216–218. [CrossRef] [PubMed]

58. Wilson, L.A.; Sawant, A.D.; Simmons, R.B.; Ahearn, D.G. Microbial contamination of contact lens storage cases and solutions. *Am. J. Ophthalmol.* **1990**, *110*, 193–198. [CrossRef]

59. Szczotka-Flynn, L.B.; Bajaksouzian, S.; Jacobs, M.R.; Rimm, A. Risk factors for contact lens bacterial contamination during continuous wear. *Optom. Vis. Sci.* **2009**, *86*, 1216–1226. [CrossRef] [PubMed]

60. Szczotka-Flynn, L.B.; Lass, J.H.; Sethi, A.; Debanne, S.; Benetz, B.A.; Albright, M.; Gillespie, B.; Kuo, J.; Jacobs, M.R.; Rimm, A. Risk factors for corneal infiltrative events during continuous wear of silicone hydrogel contact lenses. *Investig. Ophthalmol. Vis. Sci.* **2010**, *51*, 5421–5430. [CrossRef] [PubMed]

61. Szczotka-Flynn, L.; Debanne, S.M.; Cheruvu, V.K.; Long, B.; Dillehay, S.; Barr, J.; Bergenske, P.; Donshik, P.; Secor, G.; Yoakum, J. Predictive factors for corneal infiltrates with continuous wear of silicone hydrogel contact lenses. *Arch. Ophthalmol.* **2007**, *125*, 488–492. [CrossRef] [PubMed]

62. Turner, F.D.; Gower, L.A.; Stein, J.M.; Sager, D.P.; Amin, D. Compliance and contact lens care: A new assessment method. *Optom. Vis. Sci.* **1993**, *70*, 998–1004. [CrossRef] [PubMed]

63. Gray, T.B.; Cursons, R.T.; Sherwan, J.F.; Rose, P.R. Acanthamoeba, bacterial, and fungal contamination of contact lens storage cases. *Br. J. Ophthalmol.* **1995**, *79*, 601–605. [CrossRef] [PubMed]

64. Midelfart, J.; Midelfart, A.; Bevanger, L. Microbial contamination of contact lens cases among medical students. *CLAO J.* **1996**, *22*, 21–24. [PubMed]

65. Imamura, Y.; Chandra, J.; Mukherjee, P.K.; Lattif, A.A.; Szczotka-Flynn, L.B.; Pearlman, E.; Lass, J.H.; O'Donnell, K.; Ghannoum, M.A. Fusarium and Candida albicans biofilms on soft contact lenses: Model development, influence of lens type, and susceptibility to lens care solutions. *Antimicrob. Agents Chemother.* **2008**, *52*, 171–182. [CrossRef] [PubMed]

66. Baranska-Rybak, W.; Pikula, M.; Dawgul, M.; Kamysz, W.; Trzonkowski, P.; Roszkiewicz, J. Safety profile of antimicrobial peptides: Camel, citropin, protegrin, temporin a and lipopeptide on HaCaT keratinocytes. *Acta Pol. Pharm.* **2013**, *70*, 795–801. [PubMed]

67. Mannis, M.J.; Cullor, J. The use of synthetic cekropin (Shiva-11) in preservative-free timolol and contact lens solutions. *Investig. Ophthalmol. Vis. Sci.* **1993**, *34*, 859.

68. Gunshefski, L.; Mannis, M.J.; Cullor, J.; Schwab, I.R.; Jaynes, J.; Smith, W.L.; Mabry, E.; Murphy, C.J. In vitro antimicrobial activity of Shiva-11 against ocular pathogens. *Cornea* **1994**, *13*, 237–242. [CrossRef] [PubMed]

69. Sousa, L.B.; Mannis, M.J.; Schwab, I.R.; Cullor, J.; Hosotani, H.; Smith, W.; Jaynes, J. The use of synthetic Cecropin (D5C) in disinfecting contact lens solutions. *CLAO J.* **1996**, *22*, 114–117. [PubMed]

70. Fields, G.B.; Noble, R.L. Solid phase peptide synthesis utilizing 9-fluorenylmethoxycarbonyl amino acids. *Int. J. Pept. Protein Res.* **1990**, *35*, 161–214. [CrossRef] [PubMed]

71. Christensen, T. A qualitative test for monitoring coupling completeness in solid phase peptide synthesis using chloranil. *Acta Chem. Scand. B* **1979**, *33*, 763–766. [CrossRef]

72. Mrozik, W.; Markowska, A.; Guzik, L.; Kraska, B.; Kamysz, W. Determination of counter-ions in synthetic peptides by ion chromatography, capillary isotachophoresis and capillary electrophoresis. *J. Pept. Sci.* **2012**, *18*, 192–198. [CrossRef] [PubMed]

materials

MDPI

Article

In Vitro Assessment of the Antibacterial Potential of Silver Nano-Coatings on Cotton Gauzes for Prevention of Wound Infections

Federica Paladini [1], Cinzia Di Franco [2], Angelica Panico [1], Gaetano Scamarcio [2,3], Alessandro Sannino [1] and Mauro Pollini [1,*]

[1] Department of Engineering for Innovation, University of Salento, Via per Monteroni, Lecce 73100, Italy; federica.paladini@unisalento.it (F.P.); angelica.panico@unisalento.it (A.P.); alessandro.sannino@unisalento.it (A.S.)

[2] CNR-IFN U.O.S. Bari, Via Amendola 173, Bari 70126, Italy; cinzia.difranco@uniba.it (C.D.F.); gaetano.scamarcio@uniba.it (G.S.)

[3] Dipartimento Interateneo di Fisica, University of Bari Aldo Moro, Via Amendola 173, Bari 70126, Italy

* Correspondence: mauro.pollini@unisalento.it; Tel.: +39-0832-29-7562; Fax: +39-0832-29-7340

Academic Editor: Jaroslaw W. Drelich
Received: 22 February 2016; Accepted: 16 May 2016; Published: 25 May 2016

Abstract: Multidrug-resistant organisms are increasingly implicated in acute and chronic wound infections, thus compromising the chance of therapeutic options. The resistance to conventional antibiotics demonstrated by some bacterial strains has encouraged new approaches for the prevention of infections in wounds and burns, among them the use of silver compounds and nanocrystalline silver. Recently, silver wound dressings have become widely accepted in wound healing centers and are commercially available. In this work, novel antibacterial wound dressings have been developed through a silver deposition technology based on the photochemical synthesis of silver nanoparticles. The devices obtained are completely natural and the silver coatings are characterized by an excellent adhesion without the use of any binder. The silver-treated cotton gauzes were characterized through scanning electron microscopy (SEM) and thermo-gravimetric analysis (TGA) in order to verify the distribution and the dimension of the silver particles on the cotton fibers. The effectiveness of the silver-treated gauzes in reducing the bacterial growth and biofilm proliferation has been demonstrated through agar diffusion tests, bacterial enumeration test, biofilm quantification tests, fluorescence and SEM microscopy. Moreover, potential cytotoxicity of the silver coating was evaluated through 3-[4,5-dimethylthiazol-2-yl]-2,5-diphenyltetrazolium bromide colorimetric assay (MTT) and the extract method on fibroblasts and keratinocytes. Inductively coupled plasma mass spectrometry (ICP-MS) was performed in order to determine the silver release in different media and to relate the results to the biological characterization. All the results obtained were compared with plain gauzes as a negative control, as well as gauzes treated with a higher silver percentage as a positive control.

Keywords: silver dressing; biofilm; antibacterial; wound infection

1. Introduction

Wound healing can be delayed by a number of factors, such as wound colonization by microorganisms and infections. Many microorganisms in a wound produce factors detrimental to healing, such as toxins and enzymes [1]. In burn wounds, bacterial infections can frequently occur because of the accumulation of dead tissues, compromised immune system and blood supply [2,3]. In chronic wounds, bacteria persist in an adhesive matrix biofilm more resistant to antimicrobial therapy [4,5]. Multidrug-resistant organisms are increasingly implicated in both acute and chronic wound infections, thus compromising the chance of therapeutic options [6].

To prevent infection or critical colonization of wounds, the choices are locally or systemically delivered antibiotics, or dressings containing topical antimicrobial agents [7,8]. The intrinsic resistance of bacterial cell within biofilm to conventional antimicrobials has encouraged new approaches for the treatment of biofilm-associated infections; among them, the use of silver preparations represents an interesting route towards new antimicrobials. Indeed, unlike antibiotics, silver interferes with multiple components of bacterial cell structure and functions, making it less affected by specific micro-environmental variations [9–11]. Moreover, in the form of nanoparticles, silver offers many advantages in wound care and nanoparticles can also overcome existing drug resistance mechanisms [12].

The anti-inflammatory effect of nanocrystalline silver offers exciting new applications [13]. In addition to the use of silver sulfadiazine as a topical cream, more contemporary studies have explored the use of silver-impregnated dressings to manage infections [7]. Silver plays an important role in the management of burn wounds by reducing the microbial growth within a wound dressing and wound bed. The antimicrobial properties of silver-impregnated wound dressings against Gram-positive and Gram-negative bacteria have been demonstrated in several *in vivo* studies [14–16].

Burn wounds treated with silver nanoparticles demonstrated better cosmetic appearance and scarless healing [17]. In contrast to topical silver agents, wound dressings containing silver as an antimicrobial agent have been recently introduced in various designs by the wound care industry [14]. An important advantage of using silver nanoparticles in wound dressings is the continuous release of silver ions and possibility of coating the device on both the outer and inner sides, thus enhancing the antimicrobial efficacy [12].

In this work, silver-treated wound dressings have been developed for the prevention of bacterial infections in wounds and skin injuries. Traditional cotton gauzes were deposited with silver nanoparticles on both sides by adopting a deposition technology based on the *in situ* photo-reduction of silver nitrate. The efficacy of the adopted technology in providing a homogeneous coating of the cotton fibers has been verified through scanning electron microscopy (SEM) and thermo-gravimetric analysis (TGA). The antibacterial effectiveness of the silver-treated gauzes was tested in comparison with positive and negative controls. The inhibition to bacterial growth and proliferation induced by the presence of silver nanoparticles was verified through agar diffusion tests and bacterial enumeration on *Staphylococcus aureus*. The efficacy in inhibiting the bacterial biofilm formation and proliferation on the silver treated wound dressing was also demonstrated by biofilm quantification tests, fluorescence and scanning electron microscopy. The biocompatibility of the silver coating was verified through MTT assay on fibroblasts and keratinocytes and the results were related to ICP-MS analysis for evaluation of silver release in different media.

2. Experimental Section

2.1. Deposition of Silver Nanoparticles on Cotton Gauzes

Sterile cotton gauzes (10 × 10 cm, count 12/8) purchased by a local pharmacy were selected as textile substrates because of their wide use as traditional dressings in the management of wounds and burns. Coatings of silver nanoparticles were deposited on the gauzes through the *in situ* photo-reduction of silver nitrate [18] by adopting a technology described for different natural and synthetic materials [19–22]. In general, the technology involves three definite steps, namely the preparation of the impregnating silver solution, the deposition of the silver solution onto the surface of the material, and the UV exposure (365 nm) of the wet substrate. The process parameters such as the composition of the silver solution, the deposition method and the UV exposure time are defined as functions of the specific nature and application of the materials. In the specific case of cotton gauze for wound dressing application, the substrates were dip coated in an alcoholic silver solution for 1 min and then immediately exposed to ultraviolet irradiation (365 nm, 500 W, distance 20 cm) in order to induce the photo-chemical deposition of silver nanoparticles on the surface of the cotton fibers. Particularly, in this study an impregnating silver solution containing 0.5 w/v % of silver nitrate dissolved in methanol was adopted for the production of the experimental samples. Methanol was adopted as both reducing agent and solvent, even if deionized water can also be added to reduce the costs of the treatment. Moreover, samples treated with 4 w/v % Ag were also produced and adopted as positive control, whilst plain gauzes were used as negative control. After the silver treatment, the samples were carefully washed in deionized water and dried in an oven at 60 °C for 2 h.

2.2. SEM Analysis on Silver Coated Gauzes

The morphological characterization was performed by a field emission scanning electron microscope (FE-SEM), mod. ∑igma Zeiss (Jena, Germany). The samples were firstly coated with a 2 nm palladium layer by an electron beam evaporator to avoid charging and examined at a 2–15 kV acceleration voltage, 30 μm aperture, in top-view. To map the actual surface of the samples, the in-lens detector was used, revealing the secondary electrons generated in the upper range of the interaction bulb and therefore containing direct information on the sample surface morphology.

2.3. Thermo-Gravimetric Analysis TGA

The amount of silver deposited on the cotton substrates was calculated through thermo-gravimetric analysis TGA (Mettler, Columbus, OH, USA). Untreated and silver treated samples were heated from room temperature to 1000 °C in nitrogen flow and with a heating rate of 10 °C/min. The percentage of silver deposited on the gauze was calculated as the difference between the solid residues obtained from the treated and the untreated samples. Each type of sample was tested in triplicate and the results were expressed as mean values ± standard deviation.

2.4. Qualitative and Quantitative Evaluation of the Antibacterial Capability

The efficacy of the silver-treated samples in reducing the bacterial viability and proliferation was evaluated by agar diffusion tests and bacterial enumeration on *S. aureus* SA1 mucoid in comparison with the untreated sample.

The agar diffusion tests were conducted according to Standard 'SNV 195920-1992'. The procedure consisted in incubating the samples for 24 h at 37 °C in contact with bacteria on nutrient agar plates and then in evaluating the presence of an area of inhibited bacteria growth around the samples. The antibacterial capability of the sample was defined as a function of the width of the inhibition area, according to the levels provided by the Standard. Thus, if the width of the bacterial inhibition area is greater than 1 mm, a "good" antibacterial activity can be associated with the sample; on the other hand, if the sample is fully covered by bacteria, its antibacterial activity is labelled as "insufficient".

The bacterial enumeration was performed through the serial dilution method. Samples of untreated and silver treated cotton gauzes were incubated overnight at 37 °C in nutrient agar inoculated with 100 μL of *S. aureus* suspension (inoculating cell density 0.94 × 10^6 CFU/mL). After incubation, the samples were removed from the broth, and serial dilutions were performed in sterile phosphate buffered saline. From each dilution, 100 μL of the solution were extracted and plated in triplicate on nutrient agar plates. After incubation at 37 °C for 24 h, the bacterial colonies grown on the agar plates were counted and the average number of colonies was calculated for each sample. The samples were tested in triplicate. In order to evaluate the effect of the silver coating on the bacteria viability, the results were expressed as average percentages of bacterial proliferation calculated with respect to the initial concentration of the bacterial suspension.

2.5. Quantification of Bacterial Biofilm on Cotton Gauzes

Samples of untreated and silver treated cotton gauzes (1 cm^2) were placed in a 24-well microtiter and incubated overnight at 37 °C with 2 mL nutrient broth inoculated with *S. aureus* (inoculating cell density 0.94 × 10^6 CFU/mL). The samples were washed three times with phosphate-buffered saline to remove the non-adherent bacteria from the surface and then transferred to a sterile falcon containing

2 mL phosphate buffered saline. Then, the samples were vortexed for 2 min to detach the adherent bacteria and incubated at 37 °C for 4 h. For quantification of total number of bacteria, in triplicate, aliquots of suspension were analyzed with a spectrophotometer (Visible spectrophotometer V-1200, VWR, Radnor, PA, USA) and the optical density (OD600 nm) was reported as bacterial proliferation expressed in colony-forming unit (CFU/mL) (1 OD600 = 1.5×10^8 CFU/mL) [23].

Moreover, in order to evaluate the bacterial viability and to assess the bactericidal effect of the silver treated samples, the plate count method was also adopted. The aliquots of suspension were diluted in phosphate buffered saline through serial dilutions and then 100 µL of each dilution were spread on nutrient agar plates. The plates were incubated at 37 °C for 24 h and bacterial colonies grown on the agar plates were counted. The average number of colonies and the log reduction were calculated for each sample, with respect to the untreated sample.

2.6. Fluorescence Microscopy on Adherent Bacteria

Another set of samples was incubated overnight at 37 °C in a 24-well microtiter, in the same experimental conditions described in the previous section. After incubation, the samples were removed from the multi-wells, washed three times with phosphate buffered saline and transferred to other sterile wells for further SEM analysis. The wells used for gauzes incubation were washed with phosphate-buffered saline (PBS) to remove any non-adherent bacteria and then analyzed. The biofilm maturated on both gauzes and multi-wells surface was stained using green-fluorescent nucleic acid stain (SYTO9, Molecular Probes, Eugene, OR, USA). After 15 min dark incubation, bacteria were analyzed through a Zeiss inverted microscope (Axio Vert. A1 FL-LED, Oberkochen, Germany).

2.7. SEM Analysis of the Bacterial Adhesion on Cotton Gauzes

Samples of untreated and silver-coated cotton gauzes (2×2 cm, average weight 15 mg) were UV sterilized for 15 min on each side and incubated overnight at 37 °C in 3 mL of nutrient agar inoculated with 100 µL of *Staphylococcus aureus* suspension (inoculating cell density 0.94×10^6 CFU/mL). After incubation, the samples were washed with phosphate buffered saline in order to remove the non-adherent cells, whilst the adherent bacteria were fixed by using 2.5% glutaraldehyde and 2% paraformaldehyde in cacodylate buffer 0.1 M for one hour. After fixation, the samples were washed three times with cacodylate buffer for 10 min and then dehydrated in serially increasing concentrations of ethanol (25%, 50%, 75% and 100%). Each wash lasted for fifteen minutes. The samples were stored at -20 °C and conditioned at room temperature before SEM analysis.

2.8. Cytotoxicity Test

Mouse embryonic fibroblasts 3T3 and human keratinocytes HaCaT were cultured in Dulbecco's modified Eagle's medium (DMEM) supplemented with 10% fetal bovine serum (FBS) and antibiotics (1% UI/mL Streptomycin-Penicillin). Cell viability was analyzed through 3-[4,5-dimethylthiazol-2-yl]-2,5-diphenyltetrazolium bromide colorimetric assay (MTT) and the extract method, by adopting the only DMEM as positive control and the cells incubated without samples as negative control. The cells were plated at 1×10^5 cells/mL in 24-well plates and incubated for 24 h in a humidified incubator with 5% CO_2 at 37 °C. For extract preparation, 0.2 g of untreated and silver treated samples were immersed in 10 mL of phosphate buffer saline and incubated at 37 °C for 24 h. The extracts were sterilized by using 0.22 µm filters and pH close to 7.4 was measured. Then, the extracts were added to 24-wells plates with DMEM.

After incubation for 24 h, the culture medium containing the extracts was removed and 0.5 mg/mL of MTT of DMEM was added to each well and the plates were incubated in a CO_2 incubator for 2 h. After incubation, the intracellular formazan crystals were solubilized with 1 mL of isoprapanol ad centrifuged at 13,000 g for 5 min. The absorbance was determined at 550 nm using a spectrophotometer (VWR V1200). The cell viability percentage was calculated in comparison to the control group obtained without any extract. All the assays were performed in triplicate.

2.9. ICP-MS Analysis

Silver release from silver-treated cotton gauzes was calculated through inductively coupled plasma mass spectrometry ICP-MS (iCAP Q, Thermo Scientific, Waltham, MA, USA) in static conditions. Silver treated and untreated samples (0.03 g) were immersed, in triplicate, in deionized water, in phosphate-buffered saline (PBS) and in Dulbecco's Modified Eagle's Medium (DMEM) (5 mL) and incubated 24 h at 37 °C. Different media have been adopted in this characterization in order to evaluate possible interactions of the silver coating with chemical compounds, such as salts and amino acids, in comparison with deionized water. After the incubation time, an aliquot portion (25 µL) of the different media was diluted with nitric acid 1% (v/v). The samples were analysed, using silver solutions with known concentration as standards (Sigma Aldrich, St. Louis, MO, USA, Silver Standard for ICP, 1000 mg/L).

3. Results

3.1. Deposition of Silver Nanoparticles on Cotton Gauzes

The silver deposition technology based on *in situ* photo-reduction reaction has been adopted in this work to treat the experimental sample with 0.5 wt/v % silver solution and the control sample with 4 wt/v % silver solution.

3.2. SEM Analysis on Silver Coated Gauzes

The silver treated gauzes and the plain gauze were analyzed by scanning electron microscopy SEM in order to verify the distribution of the silver particles on the cotton fibers, the differences between the silver treated samples as function of the silver concentration tested and the dimension of the silver nanoparticles. The results of the SEM analysis are reported in Figure 1, where an excellent coverage of the fibers is visible for both the silver-treated samples (Figure 1b,c).

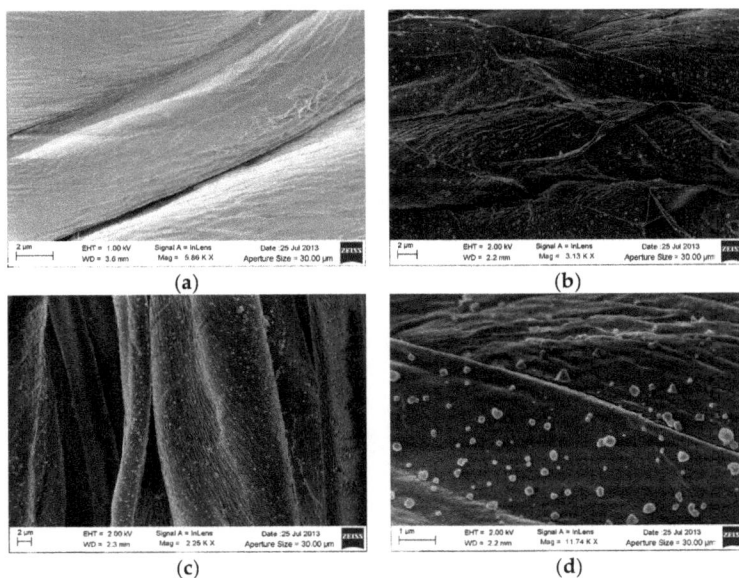

Figure 1. SEM analysis of distribution and dimension of the silver nanoparticles on the cotton fibres: (**a**) neat cotton fibers; (**b**) cotton fibers treated with 0.5 wt/v % silver; (**c**) cotton fibers treated with 4 wt/v % silver; (**d**) cotton fibers treated with 0.5 wt/v % silver at higher magnifications (×11740) for the evaluation of the dimension of the nanoparticles.

SEM analysis at higher magnification (×11740) on the sample treated with 0.5% Ag is reported in Figure 1d, where silver particles with dimensions ranging between about 100 and 300 nm can be observed.

3.3. Thermo-Gravimetric Analysis TGA

Thermo-gravimetric analysis TGA was performed in order to quantify the amount of silver deposited on the cotton gauzes as the difference between the solid residues of the silver treated sample and the untreated sample. The samples were tested in triplicate and the percentages of silver deposited resulted 0.29 ± 0.02 wt % and 2.18 ± 0.04 wt % for the sample treated with 0.5 wt/v % and 4 wt/v % respectively.

3.4. Qualitative and Quantitative Evaluation of the Antibacterial Capability

The antibacterial capability of the silver treated samples was tested on *S. aureus* SA1 mucoid through agar diffusion tests and bacterial enumeration. The results obtained by the agar diffusion tests are reported in Figure 2a–c for the untreated sample (Figure 2a), for the sample treated with 0.5 wt/v % Ag (Figure 2b) and for the sample treated with 4 wt/v % Ag (Figure 2c). As clearly visible by comparing the width of the bacterial inhibition growth areas induced by the silver treated samples, no significant difference in the antibacterial properties can be observed between the different concentrations of silver. These data were confirmed by the results obtained by the bacterial enumeration reported in Figure 3, where the percentages of bacterial proliferation calculated corresponded to 31% and 28% for sample treated with 0.5 wt/v % Ag and 4 wt/v % Ag respectively. The untreated sample resulted in a bacterial proliferation of 172%.

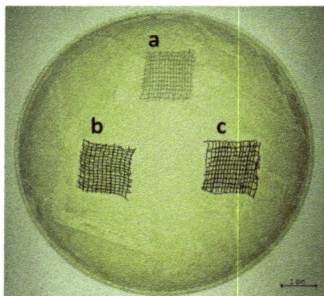

Figure 2. Agar diffusion tests on *S. aureus*: (**a**) untreated sample; (**b**) sample treated with 0.5 wt/v % silver; (**c**) sample treated with 4 wt/v % silver. The presence of the inhibition zone to bacterial growth is clearly visible around both the silver-treated samples.

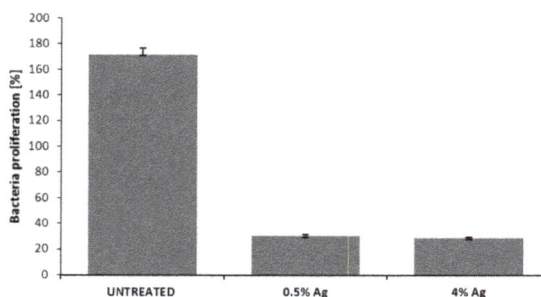

Figure 3. Bacterial proliferation expressed as percentage induced by the silver treated samples in comparison with the plain cotton gauze.

3.5. Quantification of Bacterial Biofilm on Cotton Gauzes

The bacterial biofilm was quantified through both optical density measurements and the serial dilution method. The first technique allows the estimation of total number of bacteria, while the second one is related to live cells only. The results of spectrophotometric analysis are reported in Figure 4 as bacteria proliferation, indicating a 1 log reduction in bacteria adhered to the surface of the silver treated cotton gauzes. The results obtained by the serial dilutions method are reported in Table 1. In comparison with the untreated sample, >3 log reduction was obtained by experimental sample (0.5 wt/v %), and even higher by the control sample (4 wt/v %).

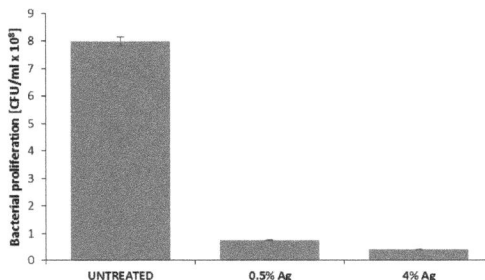

Figure 4. Quantification of bacterial biofilm on cotton gauzes through optical density measurements.

Table 1. Quantification of bacterial biofilm on cotton gauzes through serial dilution method.

Sample	CFU/mL	Log Reduction
Untreated	1.62×10^6	–
0.5% Ag	5.27×10^2	3.49
4% Ag	5.00×10	4.51

3.6. Fluorescence Microscopy on Adherent Bacteria

The results obtained by fluorescence microscopy on bacteria adhered on the multi-well surfaces are reported in Figure 5a–d for control (Figure 5a), untreated sample (Figure 5b), sample treated with 0.5% Ag (Figure 5c) and sample treated with 4% Ag (Figure 5d). Control sample refers to bacteria adhered on the tissue culture plate without any cotton sample.

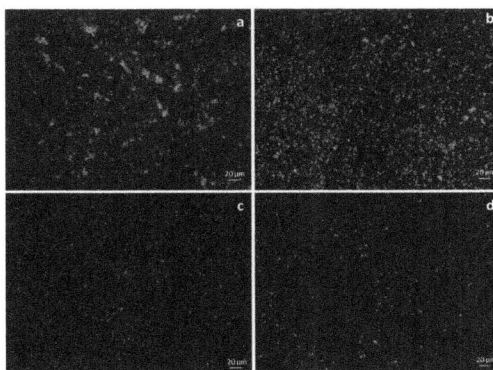

Figure 5. (**a**) Fluorescence microscopy on bacteria adhered on the multi-well plates (magnification ×40) after incubation with no sample; (**b**) untreated sample; (**c**) sample treated with 0.5 wt/v % silver; (**d**) sample treated with 4 wt/v % silver.

3.7. SEM Analysis of the Bacterial Adhesion on Cotton Gauzes

The efficacy of the silver coating in inhibiting the adhesion and the proliferation of bacterial biofilm on the cotton gauzes was analyzed through scanning electron microscopy. The SEM pictures reported in Figure 6a–c indicate the evident reduction in the adhesion of the bacterial cells to the fibers induced by the presence of silver. Indeed, a high number of *S. aureus* colonies can be observed on the neat cotton fibers (Figure 6a), whilst few isolated bacterial cells are visible on both the silver-treated samples (Figure 6b,c, arrows).

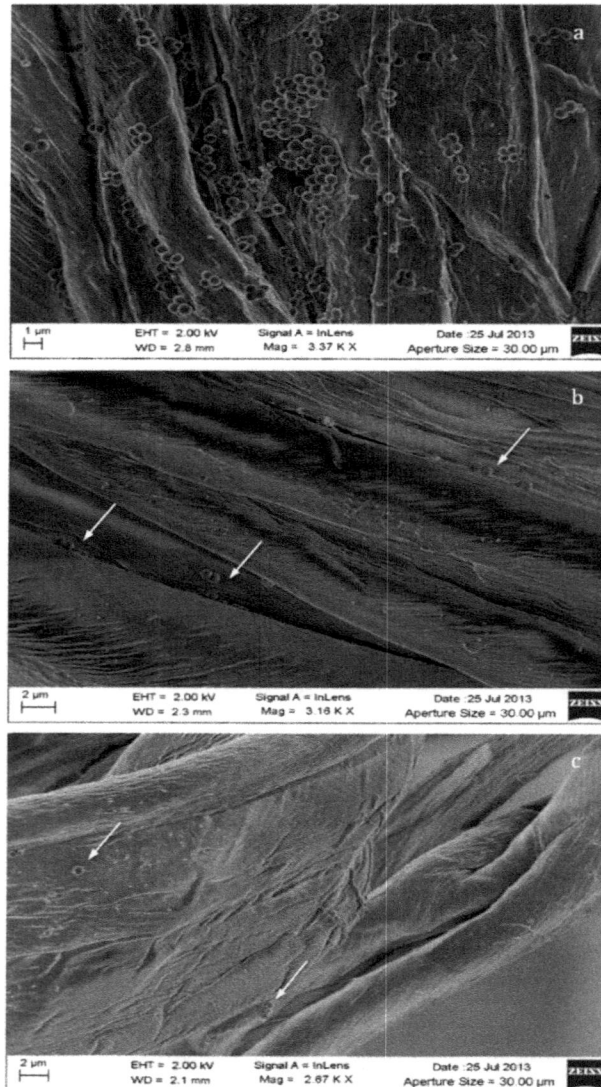

Figure 6. SEM analysis on bacterial cells adhered and proliferated on the cotton gauzes: (**a**) untreated gauze; (**b**) gauze treated with 0.5 wt/v % silver; (**c**) gauze treated with 4 wt/v % silver.

3.8. Cytotoxicity Test

Potential cytotoxicity associated to the presence of silver coating has been investigated by MTT assay through the extract method on murine fibroblasts 3T3 and human keratinocytes HaCaT. The results obtained are reported in Figure 7 as cell viability percentage, compared to cells cultured without any extract. The percentage of cell proliferation results were 103%, 105% and 21% in the presence of the extracts from untreated sample and sample treated with 0.5 wt/v % Ag and 4 wt/v % Ag respectively.

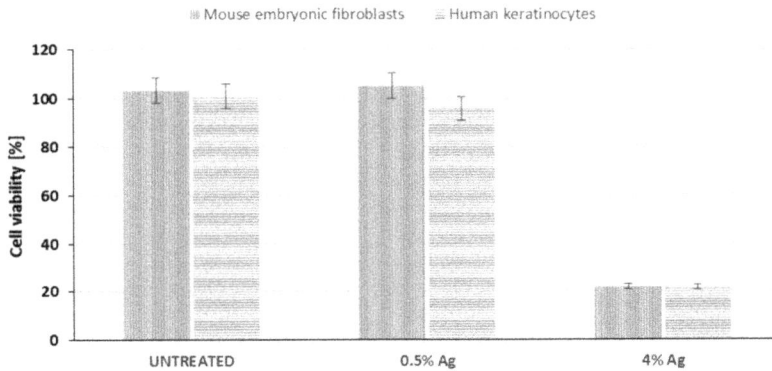

Figure 7. MTT assay for cytotoxicity evaluation.

3.9. ICP-MS Analysis

The release of silver in different media was calculated through ICP-MS analyses in static conditions, aiming to evaluate the stability of the silver coating and its possible interaction with biological fluids in terms of adhesion and cytotoxicity. The gauzes treated with 0.5 wt/v % released 0.409 ± 0.015, 0.449 ± 0.056 and 0.425 ± 0.055 ppm respectively in deionized water, in PBS and in DMEM. On the other hand, the gauzes treated with 4 wt/v % of silver released 4.505 ± 0.071, 4.160 ± 0.099 and 4.954 ± 0.014 ppm in the same media (Figure 8).

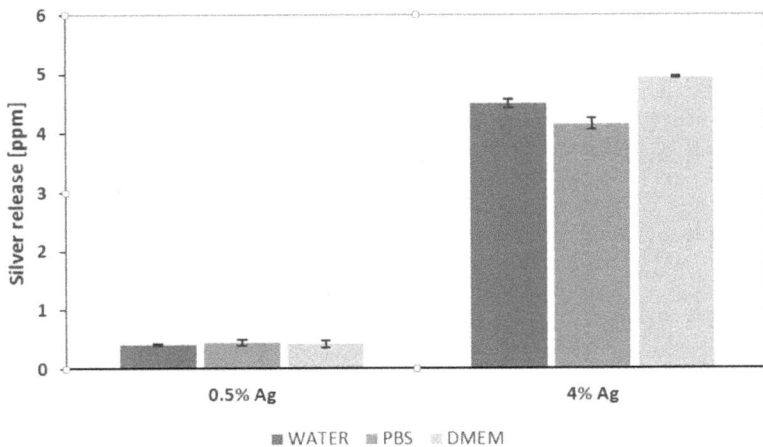

Figure 8. ICP-MS analysis performed in water, PBS and DMEM.

4. Discussion

The aim of this work was the development of silver-treated cotton gauzes and their evaluation for potential application as antibacterial dressings in wound management.

As defined in the literature, an ideal wound dressing should maintain a moist environment and oxygen permeation, should absorb excess exudates and prevent bacterial contaminations, and should be non-adherent to the wound and easily removable [24,25]. The devices developed in this work are intended to provide absorption due to the presence of cotton, and to prevent infections due to the presence of silver. However, they can also be proposed as an inherent part of a more complex device, where emollients and hydrogels can be added for improved hydration and comfort. Some wound dressings including hydrocolloids and/or other substances, have also been proposed for providing bioburden control, fast wound healing, ease of use and cost-effectiveness [26–28]. Moreover, today silver-containing dressings are widespread in the management of burn injury and acute and chronic wounds [28–30]. In addition to antimicrobial activity, silver dressings may modulate or reduce wound pain [31], and also an active role in wound healing has been associated to the presence of silver [32,33]. When compared to many wound dressings mainly based on nanocrystalline silver and silver compounds, the silver-modified gauzes presented in this work are characterized by some distinctive features related to both the nature of the materials and the production process adopted. Indeed, most of the wound dressings available today are obtained by synthetic materials, such as polyethylene, and are obtained through more complex production processes often involving the use of binders [34–38].

The main groups of silver dressings available on the market are dressings with sustained levels of silver release such as Acticoat, dressings with lower levels of silver release such as Actisorb and Aquacel, dressings with high concentration of silver at wound surface such as Contreet Foam, and dressings that release silver compounds rather than silver ions such as Urgotul [34].

In addition to the mechanism for silver release, these products are also different in terms of nature and composition. For example, Aquacel Ag is a hydrofiber-based wound dressing made of soft non-woven sodium carboxymethyl cellulose fibers integrated with ionic silver, while Urgotol Silver, developed from a lipido-colloid technology, is a non-adhesive and non-occlusive dressing made of polyester textile mesh impregnated with hydrocolloid particles and vaseline, where silver is incorporated as silver sulphate [39,40].

On the contrary, these devices are completely natural and, although the silver particles lack chemical bonds to link with natural fibers [37], an excellent adhesion of the coating to the substrate has been obtained without the use of any binder [19–22]. No complex evaporation system or expensive equipment are necessary in this silver deposition process, thus addressing an important aspect related to the scaling-up of the technology.

This paper mainly involves microbiology and cytotoxicity aspects related to the specific application of the material, aiming in particular to verify the efficacy of the silver treated wound dressings in preventing infection in skin injuries in terms of bacteria biofilm adhesion and proliferation. Thus, in the first part of the work, the silver coating was characterized through TGA and SEM analyses in order to verify the presence and distribution of the silver particles on the substrate; the second part is mainly dedicated to the biological investigation, in terms of antibacterial capability and biocompatibility. For the specific application of the textile materials object of this research work, the experimental samples were obtained by depositing cotton gauzes with 0.5 wt/v % silver solution and the results were discussed in comparison with untreated gauze and gauze treated with 4 wt/v % of silver as control samples. The presence of silver on the substrates was verified through TGA analysis and resulted different between the treated samples, as expected. Interestingly, the other results obtained by the different characterizations demonstrated that no significant differences occurred between the samples treated with 0.5 wt/v % Ag and 4 wt/v % Ag. SEM analysis (Figure 1a–d) demonstrated that 0.5 wt/v % of silver ensured a homogeneous distribution of silver nanoparticles on the cotton fibers. In order to verify and quantify the antibacterial capability of the samples, qualitative and quantitative tests, namely agar diffusion (Figure 2a–c) and bacterial enumeration tests (Figure 3), were performed on *S. aureus*, as a representative microorganism responsible for skin and wound infections. As is visible in Figure 2, both the silver concentrations exhibited inhibition areas to bacterial growth larger than 1 mm, and the bacterial enumeration tests also demonstrated an impressive reduction of the bacterial proliferation associated to samples treated with both 0.5 and 4 wt/v % silver (Figure 3). The effect of the silver coating in preventing bacteria biofilm adhesion and proliferation was also evaluated through different experiments aiming to analyze and quantify the bacterial adhesion on the cotton fibers. Textile materials can be a fertile ground for bacteria growth, and the wound site represents a particularly good environment for bacteria adhesion and proliferation [41]. The results obtained by bacterial biofilm enumeration tests demonstrated a high reduction of viability and proliferation of bacteria adhered to the surface of the device (Figure 4, Table 1), thus confirming the bacteriostatic and bactericidal effect of the silver coating. Indeed, the optical density measurements related to the total amount of bacteria demonstrated 1 log reduction between untreated and silver treated samples, while the evaluation of bacteria viability through the serial dilutions method indicated log reduction >3, thus confirming the bactericidal effect of deposited silver [42,43]. Moreover, fluorescence microscopy performed on bacteria stained on the multi-well plate after incubation with and without the samples (Figure 5) indicated that the number of bacteria was significantly reduced in presence of the silver coatings (Figure 5c,d), in comparison with the untreated sample (Figure 5b) and control (Figure 5a). As expected, a high concentration of bacteria organized in clusters of colonies was observed in control sample, *i.e.*, in case of only medium inoculated with bacteria. On the other hand, compared to control, the untreated sample (Figure 5b) resulted in a lower concentration of bacteria adhered on the multi-well plate, because of the partial adhesion of bacteria also on the gauze surface. At this purpose, SEM analysis was also performed on bacteria biofilm grown on the textile substrates (Figure 6a–c), confirming the effectiveness of the silver coatings in reducing the bacteria adhesion and proliferation on the gauze. Indeed, while a few isolated bacterial cells can be observed on the silver-treated samples (Figure 6b,c), a large number of bacterial colonies aggregated in bacterial communities can be observed on the untreated sample, indicating an initial step of biofilm growth (Figure 6a). The presence of silver coating successfully inhibited the bacteria growth, and significant differences were not observed between the silver concentrations tested. All the results indicated that the silver coating obtained by 0.5 wt/v % Ag solution was as effective as 4 wt/v % Ag against *S. aureus*, and that it successfully inhibited the bacterial colonization and biofilm formation on the dressing. In the graph reported in Figure 7, no significant differences can be observed in percentage of cell proliferation between untreated and 0.5 wt/v % Ag-treated samples. These data confirmed no cytotoxic effect of silver treated textile substrates and no skin irritation or hypoallergenicity effects through *in vivo* testing on selected patients [44–47]. Although the cell viability was reduced by the samples treated with 4 wt/v % silver, the highest values of silver release produced results that were still lower than the limit of cytotoxicity reported in the literature [48–50]. On the other hand, the presence of the silver coating on the cotton gauze treated with 0.5 wt/v % did not affect the cell viability. ICP-MS analysis performed in different media demonstrated the strong adhesion of the coating to the cotton substrate, thus also confirming that very low amounts of silver are released by the experimental samples even in simulated physiological conditions. The experiments were carried out in deionized water as control, in PBS in order to reproduce physiological conditions, and in DMEM in order to relate the data to cytotoxicity tests (Figure 8). As expected, samples treated with 4 wt/v % Ag released values of silver higher than samples treated with 0.5 wt/v %, and the results were consistent with the biological characterization. Moreover, the presence of a biological environment did not affect the silver release, thus confirming the stability of the coating and the perfect adhesion to the textile substrate.

5. Conclusions

The aim of this work was the development of effective and low-cost silver dressings for the prevention of bacterial infections in wound, burns and skin injuries. Conventional cotton gauzes extensively used in wound care for their absorbent properties and economic features were deposited with silver nanoparticles through the *in situ* photo-reduction of silver nitrate. A low content of silver (0.5 wt/v %) was adopted and tested in comparison with a higher silver percentage (4 wt/v %) and an untreated sample. TGA and SEM analysis demonstrated the presence of nanoparticles and their good distribution on the treated samples. The microbiological characterizations demonstrated that the gauze treated with 0.5 wt/v % of silver was as effective as that treated with 4 wt/v % of silver against bacteria in terms of viability and biofilm adhesion and growth. Due to the good antimicrobial properties and biocompatibility demonstrated, the textile materials developed can be considered a promising alternative to conventional wound dressings, with clear advantages in terms of prevention of infections and costs.

Acknowledgments: The authors acknowledge the Ministry of University and Scientific Research (MIUR) of Italy, PON program 2007–2013 for financial support.

Author Contributions: In this research work, Federica Paladini and Mauro Pollini conceived and designed the experiments and wrote the article; Cinzia Di Franco and Gaetano Scamarcio provided data and discussion about SEM analysis; Angelica Panico performed the microbiological characterizations and Alessandro Sannino analyzed data and revisions. All the authors discussed the results, revised and approved the final version of the manuscript.

Conflicts of Interest: The authors declare no conflict of interest.

Abbreviations

The following abbreviations are used in this manuscript:

SEM	Scanning Electron Microscopy
TGA	Thermo-Gravimetric Analysis
MTT	3-[4,5-dimethylthiazol-2-yl]-2,5-diphenyltetrazolium bromide colorimetric assay
ICP-MS	Inductively Coupled Plasma Mass Spectrometry

References

1. Percival, S.; Slone, W.; Linton, S.; Okel, T.; Corum, L.; Thomas, J.G. The antimicrobial efficacy of a silver alginate dressing against a broad spectrum of clinically relevant wound isolates. *Int. Wound J.* **2011**, *8*, 237–243. [CrossRef] [PubMed]

2. Bloemsma, G.C.; Dokter, J.; Boxma, H.; Oen, I.M. Mortality and causes of death in a burn center. *Burns* **2008**, *34*, 1103–1107. [CrossRef] [PubMed]

3. Abedini, F.; Ahmadi, A.; Yavari, A.; Hosseini, V.; Mousavi, S. Comparison of silver nylon wound dressing and silver sulfadiazine in partial burn wound therapy. *Int. Wound J.* **2013**, *10*, 573–578. [CrossRef] [PubMed]

4. Rhoads, D.D.; Wolcott, R.D.; Percival, S.L. Biofilms in wounds: Management strategies. *J. Wound Care* **2008**, *17*, 502–508. [CrossRef] [PubMed]

5. Lipsky, B.A.; Hoey, C. Topical antimicrobial therapy for treating chronic wounds. *Clin. Infect. Dis.* **2009**, *49*, 1541–1549. [CrossRef] [PubMed]

6. Bowler, P.G.; Welsby, S.; Towers, V.; Booth, R.; Hogarth, A.; Rowlands, V.; Joseph, A.; Jones, S.A. Multidrug-resistant organisms, wounds and topical antimicrobial protection. *Int. Wound J.* **2012**, *9*, 387–396. [CrossRef] [PubMed]

7. Carter, M.J.; Tingley-Kelley, K.; Warriner, R.A. Silver treatments and silver-impregnated dressings for the healing of leg wounds and ulcers: A systematic review and meta-analysis. *J. Am. Acad. Dermatol.* **2010**, *63*, 668–679. [CrossRef] [PubMed]

8. Cutting, K.; White, R.; Hoekstra, H. Topical silver-impregnated dressings and the importance of the dressing technology. *Int. Wound J.* **2009**, *6*, 396–402. [CrossRef] [PubMed]

9. Bjarnsholt, T.; Kirketerp-Moller, K.; Kristiansen, S.; Phipps, R.; Nielsen, A.K.; Jensen, P.O.; Hoiby, N.; Givskov, M. Silver against *Pseudomonas aeruginosa* biofilms. *APMIS* **2007**, *115*, 921–928. [CrossRef] [PubMed]

10. Lansdown, A.B. Silver I: Its antibacterial properties and mechanism of action. *J. Wound Care* **2002**, *11*, 125–130. [CrossRef] [PubMed]

11. Kostenko, V.; Lyczak, J.; Turner, K.; Martinuzzi, R.J. Impact of Silver-Containing Wound Dressings on Bacterial Biofilm Viability and Susceptibility to Antibiotics during Prolonged Treatment. *Antimicrob. Agents Chemother.* **2010**, *54*, 5120–5131. [CrossRef] [PubMed]

12. Pelgrift, R.Y.; Friedman, A.J. Nanotechnology as a therapeutic tool to combat microbial resistance. *Adv. Drug Deliv. Rev.* **2013**, *65*, 1803–1815. [CrossRef] [PubMed]

13. Edwards-Jones, V. The benefits of silver in hygiene, personal care and healthcare. *Lett. Appl. Microbiol.* **2009**, *49*, 147–152. [CrossRef] [PubMed]

14. Castellano, J.J.; Shafii, S.M.; Ko, F.; Donate, G.; Wright, T.E.; Mannari, R.J.; Payne, W.G.; Smith, D.J.; Robson, M.C. Comparative evaluation of silver-containing antimicrobial dressings and drugs. *Int. Wound J.* **2007**, *4*, 114–122. [CrossRef] [PubMed]

15. Cavanagh, M.H.; Burrell, R.E.; Nadworny, P.L. Evaluating antimicrobial efficacy of new commercially available silver dressings. *Int. Wound J.* **2010**, *7*, 394–405. [CrossRef] [PubMed]

16. Percival, S.L.; Thomas, J.G.; Slone, W.; Linton, S.; Corum, L.; Okel, T. The efficacy of silver dressings and antibiotics on MRSA and MSSA isolated from burn patients. *Wound Repair Regen.* **2011**, *19*, 767–774. [CrossRef] [PubMed]

17. Rai, M.; Yadav, A.; Gade, A. Silver nanoparticles as a new generation of antimicrobials. *Biotechnol. Adv.* **2009**, *27*, 76–83. [CrossRef] [PubMed]

18. Pollini, M.; Sannino, A.; Maffezzoli, A.; Licciulli, A. Antibacterial Surface Treatments Based on Silver Clusters Deposition. U.S. Patent 20090130181, 21 May 2009.

19. Pollini, M.; Paladini, F.; Licciulli, A.; Maffezzoli, A.; Sannino, A.; Nicolais, L. Antibacterial natural leather for application in the public transport system. *J. Coat. Technol. Res.* **2013**, *10*, 239–245. [CrossRef]

20. Pollini, M.; Paladini, F.; Licciulli, A.; Maffezzoli, A.; Sannino, A. Engineering Nanostructured Silver Coatings for Antimicrobial Applications. In *Nanoantimicrobials Progress and Prospects*; Cioffi, N., Rai, M., Eds.; Springer: Heidelberg, Germany; Dordrecht, The Netherlands; London, UK; New York, NY, USA, 2012; pp. 313–336.

21. Paladini, F.; Cooper, I.R.; Pollini, M. Development of antibacterial and antifungal silver-coated polyurethane foams as air filtration units for the prevention of respiratory diseases. *J. Appl. Microbiol.* **2014**, *116*, 710–717. [CrossRef] [PubMed]

22. Pollini, M.; Paladini, F.; Sannino, A.; Picca, R.A.; Sportelli, M.C.; Cioffi, N.; Nitti, M.A.; Valentini, M.; Valentini, A. Nonconventional Routes to Silver Nanoantimicrobials: Technological Issues, Bioactivity and Applications. In *Nanotechnology in Diagnosis, Treatment and Prophylaxis of Infectious Diseases*; Rai, M., Kon, K., Eds.; Elsevier: London, UK; San Diego, CA, USA; Waltham, MA, USA; Oxford, UK, 2015; pp. 87–105.

23. Chang, Y.C.; Yang, C.Y.; Sun, R.L.; Cheng, Y.F.; Kao, W.C.; Yang, P.C. Rapid single cell detection of *Staphylococcus aureus* by aptamer-conjugated gold nanoparticles. *Sci. Rep.* **2013**, *3*. [CrossRef] [PubMed]

24. Napavichayanun, S.; Amornsudthiwat, P.; Pienpinijtham, P.; Aramwit, P. Interaction and effectiveness of antimicrobials along with healing-promoting agents in a novel biocellulose wound dressing. *Mater. Sci. Eng. C* **2015**, *55*, 95–104. [CrossRef] [PubMed]

25. Calamak, S.; Erdogdu, C.; Ozalp, M.; Ulubayram, K. Silk fibroin based antibacterial bionanotextiles as wound dressing materials. *Mater. Sci. Eng. C* **2014**, *43*, 11–20. [CrossRef] [PubMed]

26. Khundkar, R.; Malic, C.; Burge, T. Use of Acticoat™ dressing in burns: What is the evidence? *Burns* **2010**, *36*, 751–758. [CrossRef] [PubMed]

27. Verbelen, J.; Hoeksema, H.; Heyneman, A.; Pirayesh, A.; Monstrey, S. Aquacel® Ag dressing *versus* Acticoat™ dressing in partial thickness burns: A prospective, randomized, controlled study in 100 patients. Part 1: Burn wound healing. *Burns* **2013**. [CrossRef] [PubMed]

28. Huang, L.; Dai, T.; Xuan, Y.; Tegos, G.P.; Hamblin, M.R. Synergistic combination of chitosan acetate with nanoparticle silver as a topical antimicrobial: Efficacy against bacterial burn infections. *Antimicrob. Agents Chemother.* **2011**, *55*, 3432–3438. [CrossRef] [PubMed]

29. Roman, M.; Rigo, C.; Munivrana, I.; Vindigni, V.; Azzena, B.; Barbante, C.; Fenzi, F.; Guerriero, P.; Cairns, W.R. Development and application of methods for the determination of silver in polymeric dressings used for the care of burns. *Talanta* **2013**, *115*, 94–103. [CrossRef] [PubMed]

30. Knetsch, M.L.W.; Koole, L.H. New Strategies in the Development of Antimicrobial Coatings: The Example of Increasing Usage of Silver and Silver Nanoparticles. *Polymers* **2011**, *3*, 340–366. [CrossRef]

31. Abboud, E.C.; Legare, T.B.; Settle, J.C.; Boubekri, A.M.; Barillo, D.J.; Marcet, J.E.; Sanchez, J.E. Do silver-based wound dressings reduce pain? A prospective study and review of the literature. *Burns* **2014**, *40*. [CrossRef] [PubMed]

32. Fong, J.; Wood, F. Nanocrystalline silver dressings in wound management: A review. *Int. J. Nanomed.* **2006**, *1*, 441–449. [CrossRef]

33. Lin, Y.H.; Hsu, W.S.; Chung, W.Y.; Ko, T.H.; Lin, J.H. Silver-based wound dressings reduce bacterial burden and promote wound healing. *Int. Wound J.* **2015**. [CrossRef] [PubMed]

34. Leaper, D.J. Silver dressings: Their role in wound management. *Int. Wound J.* **2006**, *3*, 282–294. [CrossRef] [PubMed]

35. Ip, M.; Lui, S.L.; Poon, V.K.; Lung, I.; Burd, A. Antimicrobial activities of silver dressings: An *in vitro* comparison. *J. Med. Microbiol.* **2006**, *55*, 59–63. [CrossRef] [PubMed]

36. Moiemen, N.S.; Shale, E.; Drysdale, K.J.; Smith, G.; Wilson, Y.T.; Papini, R. Acticoat dressings and major burns: Systemic silver absorption. *Burns* **2011**, *37*, 27–35. [CrossRef] [PubMed]

37. El-Rafie, M.H.; Ahmed, H.B.; Zahran, M.K. Characterization of nanosilver coated cotton fabrics and evaluation of its antibacterial efficacy. *Carbohydr. Polym.* **2014**, *107*, 74–181. [CrossRef] [PubMed]

38. Cortese, B.; Caschera, D.; Padeletti, G.; Ingo, G.M.; Gigli, G. A brief review of surface-functionalized cotton fabrics. *Surf. Innov.* **2013**, *1*, 140–156. [CrossRef]

39. Barnea, Y.; Weiss, J.; Gur, E. A review of the applications of the hydrofiber dressing with silver (Aquacel Ag) in wound care. *Ther. Clin. Risk Manag.* **2010**, *6*, 21–27. [CrossRef] [PubMed]

40. Harding, K.; Gottrup, F.; Jawień, A.; Mikosiński, J.; Twardowska-Saucha, K.; Kaczmarek, S.; Sopata, M.; Shearman, C.; Pieronne, A.; Kommala, D. A prospective, multi-centre, randomised, open label, parallel, comparative study to evaluate effects of AQUACEL® Ag and Urgotul® Silver dressing on healing of chronic venous leg ulcers. *Int. Wound J.* **2012**, *9*, 285–294. [CrossRef] [PubMed]

41. Tang, B.; Kaur, J.; Sun, L.; Wang, X. Multifunctionalization of cotton through *in situ* green synthesis of silver nanoparticles. *Cellulose* **2013**, *20*, 3053–3065. [CrossRef]

42. Basri, D.F.; Xian, L.W.; Shukor, N.I.A.S.; Latip, J. Bacteriostatic Antimicrobial Combination: Antagonistic Interaction between Epsilon-Viniferin and Vancomycin against Methicillin-Resistant *Staphylococcus aureus*. *Biomed. Res. Int.* **2014**, *2014*. [CrossRef] [PubMed]

43. Lee, G.C.; Burgess, D.S. Polymyxins and Doripenem Combination Against KPC-Producing Klebsiella pneumonia. *J. Clin. Med. Res.* **2013**, *5*, 97–100. [PubMed]

44. Paladini, F.; De Simone, S.; Sannino, A.; Pollini, M. Antibacterial and Antifungal Dressings Obtained by Photochemical Deposition of Silver Nanoparticles. *J. Appl. Polym. Sci.* **2014**, *131*. [CrossRef]

45. Paladini, F.; Picca, R.A.; Sportelli, M.C.; Cioffi, N.; Sannino, A.; Pollini, M. Surface chemical and biological characterization of flax fabrics modified with silver nanoparticles for biomedical applications. *Mater. Sci. Eng. C Mater. Biol. Appl.* **2015**, *52*, 1–10. [CrossRef] [PubMed]

46. Paladini, F.; Sannino, A.; Pollini, M. *In vivo* testing of silver treated fibers for the evaluation of skin irritation effect and hypoallergenicity. *J. Biomed. Mater. Res. B Appl. Biomater.* **2014**, *102*, 1031–1037. [CrossRef] [PubMed]

47. Paladini, F.; Pollini, M.; Sannino, A.; Ambrosio, L. Metal-Based Antibacterial Substrates for Biomedical Applications. *Biomacromolecules* **2015**, *16*, 1873–1885. [CrossRef] [PubMed]

48. Schierholz, J.M.; Lucasj, L.J.; Rump, A.; Pulverer, G. Efficacy of silver-coated medical devices. *J. Hosp. Infect.* **1998**, *40*, 257–262. [CrossRef]

49. Damm, C.; Munstedt, H.; Rosch, A. The antimicrobial efficacy of polyamide 6/silver-nano- and microcomposites. *Mater. Chem. Phys.* **2008**, *108*, 61–66. [CrossRef]

50. Greulich, C.; Braun, D.; Peetsch, A.; Diendorf, J.; Siebers, B.; Epple, M.; Koller, M. The toxic effect of silver ions and silver nanoparticles towards bacteria and human cells occurs in the same concentration range. *RSC Adv.* **2012**, *2*, 6981–6987. [CrossRef]

materials

MDPI

Article

Antimicrobial Properties of Diamond-Like Carbon/Silver Nanocomposite Thin Films Deposited on Textiles: Towards Smart Bandages

Tadas Juknius [1], Modestas Ružauskas [2], Tomas Tamulevičius [1,3,*], Rita Šiugždinienė [2], Indrė Juknienė [2], Andrius Vasiliauskas [2], Aušrinė Jurkevičiūtė [1] and Sigitas Tamulevičius [1,3]

[1] Institute of Materials Science, Kaunas University of Technology, K. Baršausko St. 59, 51423 Kaunas, Lithuania; Tadas.Juknius@ktu.edu (T.J.); Ausrine.Jurkeviciute@ktu.lt (A.J.); Sigitas.Tamulevicius@ktu.lt (S.T.)
[2] Veterinary Academy, Lithuanian University of Health Sciences, Tilžės St. 18, 47181 Kaunas, Lithuania; Modestas.Ruzauskas@lsmuni.lt (M.R.); Rita.Siugzdiniene@lsmuni.lt (R.Š.); Indre.Jukniene@fc.lsmuni.lt (I.J.); Andrius.Vasiliauskas@ktu.lt (A.V.)
[3] Department of Physics, Kaunas University of Technology, Studentų St. 50, 51368 Kaunas, Lithuania
* Correspondence: Tomas.Tamulevicius@ktu.lt; Tel.: +370-37-313-432

Academic Editor: Mauro Pollini
Received: 31 March 2016; Accepted: 10 May 2016; Published: 13 May 2016

Abstract: In the current work, a new antibacterial bandage was proposed where diamond-like carbon with silver nanoparticle (DLC:Ag)-coated synthetic silk tissue was used as a building block. The DLC:Ag structure, the dimensions of nanoparticles, the silver concentration and the silver ion release were studied systematically employing scanning electron microscopy, energy dispersive X-ray spectroscopy and atomic absorption spectroscopy, respectively. Antimicrobial properties were investigated using microbiological tests (disk diffusion method and spread-plate technique). The DLC:Ag layer was stabilized on the surface of the bandage using a thin layer of medical grade gelatin and cellulose. Four different strains of *Staphylococcus aureus* extracted from humans' and animals' infected wounds were used. It is demonstrated that the efficiency of the Ag^+ ion release to the aqueous media can be increased by further RF oxygen plasma etching of the nanocomposite. It was obtained that the best antibacterial properties were demonstrated by the plasma-processed DLC:Ag layer having a 3.12 at % Ag surface concentration with the dominating linear dimensions of nanoparticles being 23.7 nm. An extra protective layer made from cellulose and gelatin with agar contributed to the accumulation and efficient release of silver ions to the aqueous media, increasing bandage antimicrobial efficiency up to 50% as compared to the single DLC:Ag layer on textile.

Keywords: nanocomposite; silver; bandage; antimicrobial; *S. aureus*

PACS: 81.07.-b; AGRICOLASCC: L832

1. Introduction

Quite often, due to bad care of a wound (or a weak immune system), pathogenic microorganisms cause wound inflammation. Bacteria, like methicillin-resistant *Staphylococcus aureus* (MRSA), can cause severe infections with bad prognosis and consequences [1–3]. On the other hand, small wounds after injuries can be healed using simple cotton bandages, which usually do not protect the damaged tissues from bacteria and secondary infection.

It is known that bacterial adhesion is the first step in colonization of wounded skin and the formation of a biofilm [4]. Therefore, it is very important to have the antibacterial surface directly on the bandage to avoid the formation of these bacteria films, preventing bacteria multiplication inside the bandage [5]. Bacterial infections are usually treated with antibiotics; however, a good effect is not obtained in every case, and this is because many bacteria are immunized to antibiotics after being subjected to the treatment [6]. Moreover, antimicrobial resistance is directly associated with over-usage of antibiotics; thus, alternative methods for treatment are necessary. One of the most important pathogenic bacteria associated with different types of infections, including skin and soft tissues, is *S. aureus* [7]. Infections caused by this species sometimes are very hard to treat because of its multiple-resistance. Due to this reason, *S. aureus* is reputed as one of the most intractable pathogenic bacteria in the history of antibiotic chemotherapy [8].

The closure of wounds and fistulas using silver sutures was shown to be very successful in preventing infections [9]. Silver can be applied in different forms, namely as a metal, as a compound or as a free dissolved ion. The famous Hippocrates first described silver's antimicrobial properties in 400 BC [10]. The ancient users of silver had no idea what form of silver worked best, but they just observed the positive effects of silver and silver salts. They also realized that silver worked best when some moisture was present. Now, it is proven that the silver ions (Ag^+) are responsible for the antimicrobial activity [11]. It is known that silver is an antiseptic metal and can act against bacteria in different ways due to Ag cation release in aqueous media. Ag^+ can destruct the cell membranes, destroy the respiratory enzyme system or can block DNA replication. This variability of antibacterial mechanisms of Ag hinders bacterial resistance formation [12]. The possible antibacterial action mechanisms of silver nanoparticles (Ag NPs) are explained in the following way. Ag^+ ions are supposed to bind to sulfhydryl groups, which lead to protein denaturation by reducing disulfide bonds; Ag^+ can complex with electron donor groups containing sulfur, oxygen or nitrogen that are normally present as thiols or phosphates on amino acids and nucleic acids [13]. Ag NPs react with sulfur-rich proteins in the bacteria cell membrane and the interior of the cell or with phosphorous-containing compounds, such as DNA. Accordingly, the morphological changes in the bacteria cell membrane and the possible damage of DNA caused by the reaction with Ag NPs disturb the respiratory chain or cell division processes, leading to cell death [14]. The antibacterial activity of a zero-valent silver phase strictly depends on the surface development of the solid, since silver atoms/ions required to accomplish the antibacterial activity are released only from the surface. Consequently, when this solid phase is in a powdered form, the resulting antibacterial activity can be significantly increased, and an ultrafine silver powder may result in several orders of magnitude more activity than the corresponding bulk solid [15]. The Ag NPs are known to oxidize to Ag ions when they interact with water molecules. Ag nanoparticles could provide sustained release of sufficient Ag ion compared to Ag salts and bulk metallic forms due to higher active surface to volume ratios [16]. Finally, the nanosized silver antibacterial activity is enhanced for two main reasons: the fraction of surface silver atoms increases with decreasing particle size, and a greater fraction of surface silver atoms is weakly bonded to the particle surface, which can be easily released to the surrounding medium. It is agreed that the Ag NP nanocomposite antimicrobial activity is basically related to its ability to release Ag^+ over time [15].

The amount of biotechnological products containing silver nanoparticles in their composition is increasing every day. For example, they are being used in the impregnation of wound dressings, medical devices, dental materials, fabrics, among others, besides the combination with antibiotics for the observation of a synergistic antibacterial effect [13]. Because of the increased attention to antibacterial nanocomposites in the last few decades, there is significant interest in studying Ag ion release from Ag NPs to optimize the nanocomposite performance and to reduce the negative effects on human cells and the environment [17]. Ag nanoparticles have to be placed in media or on a surface, which could have direct contact with skin tissues [18]. As an example, synthetic silk can be rinsed in colloidal silver solution and, later on, dried out. However, in such a case, due to rinsing, less Ag nanoparticles remain, and finally, the surface is polluted with the remainder of the chemicals used in the process [19]. Therefore, this technology is not suitable for bandages. Antibacterial surfaces can be produced, as well as a coating, where Ag-containing coatings can be envisaged as a pure Ag layer, or Ag nanoparticles embedded in a matrix [12].

In this case, diamond-like carbon (DLC) can be considered as a candidate, as it is known as a versatile coating material that finds a variety of biomedical applications, including endoprosthesis and dental implants. It provides mechanical robustness and cell-compatibility at the same time. To broaden this range of beneficial properties even more, trials were reported when silver or silver nanoparticles were embedded in order to add antibacterial properties [4,20].

Considering the final product, it should be mentioned that now, some bandages with silver-coated textile are available on the market [21–23], where a thin silver layer is used that may release silver ions if moisture is present. Some manufacturers offer hydrogel-based wound dressings [24,25], where the hydrogel material is saturated with silver ions. In addition, hydrogel contains water and crosslinked molecules, the structure of which ensures moisture absorption [26].

In the current work, DLC:Ag nanocomposite films were deposited on textile (synthetic silk) as a part of a smart bandage prototype demonstrating antibacterial properties. The antibacterial properties of such nanocomposites were systematically tested against *S. aureus* emphasizing conditions of DC magnetron sputtering and, further, plasma processing steps. DLC:Ag properties were characterized employing scanning electron microscopy (SEM) with energy dispersive X-ray spectroscopy (EDS) and atomic absorption spectroscopy (AAS). Further development of the smart bandage by integrating new functionalities is in progress and will be published in the forthcoming publications.

2. Results

2.1. Structure and Composition of DLC:Ag Layers Deposited on Textile and Crystalline Silicon

From the EDS measurements, it was obtained that the chemical content of Ag in as-deposited DLC:Ag films varies from 0.46–6.43 at % (Table 1).

Table 1. Averaged DLC:Ag chemical composition obtained with EDS.

Sample No.	Carbon (at %)	Silver (at %)	Oxygen (at %)
1	67.54 ± 1.33 [1]	0.46 ± 0.02	32.01 ± 1.33
2	61.55 ± 0.58	3.12 ± 0.11	31.16 ± 4.65
3	48.92 ± 3.65	6.43 ± 0.65	44.65 ± 4.27

[1] Uncertainty is one standard deviation.

SEM measurements revealed that the lowest Ag content films do not have visible Ag nanoparticles on the surface (Sample 1). A few particles were visible only on the surface of the silk substrates (see Figure 1a) that were not present prior to deposition. The middle and highest investigated Ag concentration samples (Samples 2 and 3) had a number of silver clusters on the surface (Figure 1b,c). SEM micrographs and NP analysis are depicted in Figure 1.

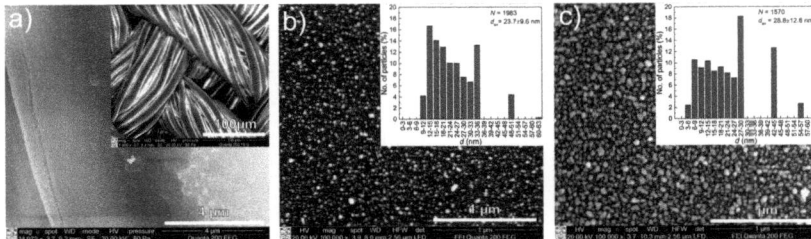

Figure 1. SEM micrographs of DLC:Ag films of different Ag content on different substrates: (**a**) lowest Ag content film (Sample 1; scale bar: 4 μm) on silk; the inset depicts the silk structure (scale bar: 100 μm); (**b**,**c**) summarize the information about the medium and highest Ag concentration film (Samples 2 and 3, respectively; scale bar: 1 μm) on silicon substrates; insets depict the particle size distribution number of analyzed particles (N) and the average particle diameters (d_{av}) of the corresponding SEM micrographs.

2.2. Antimicrobial Activity of Virgin and RF O₂ Plasma Processed DLC:Ag Films

Three as-deposited and O_2 plasma-processed DLC:Ag thin films containing different amounts of Ag 0.5–6.0 at % deposited on synthetic silk were investigated employing the disk diffusion method. Figure 2 demonstrates the results of antimicrobial activity for Samples 1–3.

Figure 2. Testing of antimicrobial properties with the *S. aureus* LTSaM01 strain bacteria (disk diffusion method) of virgin Samples 1 (**a**); 2 (**b**) and 3 (**c**); as well as after 20-s plasma processing ((**d–f**) respectively). The clear zone in (**d–f**) indicates areas where no bacteria multiplication is observed.

One can see that the virgin DLC:Ag-coated sample does not show an expressed bacteria inhibition area for the *S. aureus* LTSaM01 strain (the same results were found for all investigated strains around the samples in the Petri dishes), while the O_2 plasma-processed sample demonstrates up to a 2.5-mm inhibition zone.

The inhibition zones' dimensions measured with four different strains after different processing duration with O_2 plasma for three different Ag concentration samples on silk are depicted in Figure 3.

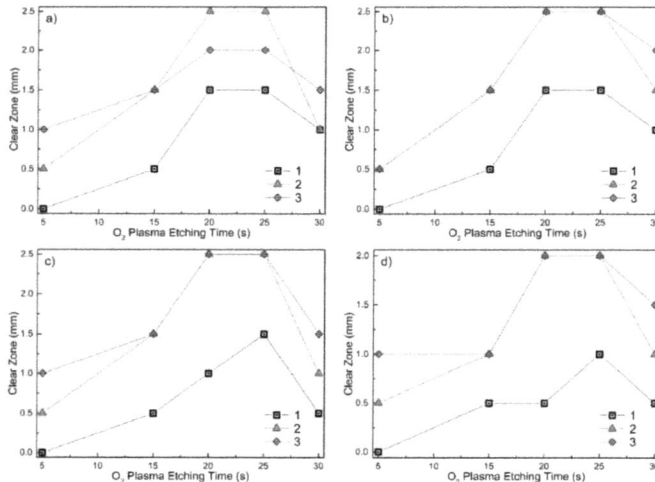

Figure 3. Antimicrobial effect (linear dimensions of the clear zone) for three different Ag concentration DLC:Ag films deposited on silicon and processed with O_2 plasma for different durations (5–30 s) (Samples 1–3) were tested with four different *S. aureus* bacteria strains, (**a**) LTSaDA01; (**b**) LTSaM01; (**c**) LTSa635 and (**d**) LTSa603, the employing disk diffusion method. Measurement uncertainty: 0.5 mm.

From Figure 3, one can see that after 5 s of O_2 plasma etching, Sample 1 showed no antimicrobial effect with all investigated bacteria strains. Sample 2 had only a weak effect for all bacteria strains, *i.e.*, the inhibition zone was ⩽0.5 mm (see Figure 3). Sample 3 with 6.43 at % of silver has shown even better results with the shortest plasma processing duration, *i.e.*, the inhibition zone was 1 mm, except bacteria LTSaM01 (see Figure 3b), where a clear zone of 0.5 mm was obtained. After 15 s of etching in O_2 plasma, Sample 1 had a visible antimicrobial effect for all bacteria strains; the inhibition zone was 0.5 mm. Samples 2 and 3 demonstrated a larger inhibition zone: 1.5 mm; except strain LTSa603 (see Figure 3d), where the inhibition zone was 1 mm. The best results in terms of inhibition zone were observed for the samples etched for 20 s and 25 s. As can be seen from Figure 2d–f and Figure 3, plasma etching strongly increased the inhibition effect for all four *S. aureus* strains for all investigated samples with different Ag content nanocomposite DLC:Ag films.

The antimicrobial effect of differently plasma-processed DLC:Ag nanocomposite films on the investigated strains could be related to the surface morphology changes and opening of the Ag NPs' surface from the DLC matrix (see Figure 4). The chemical composition of the films after different O_2 plasma processing, providing the characteristic weak and strong antimicrobial effects on the investigated bacteria strains, is depicted in Figure 5.

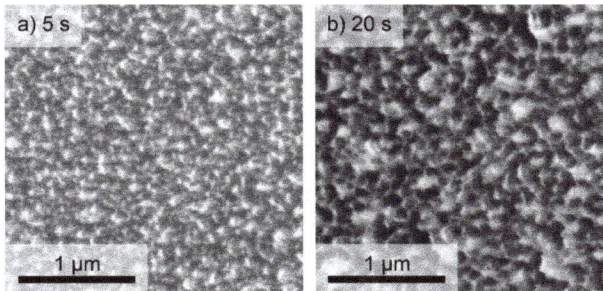

Figure 4. SEM micrographs of DLC:Ag (Sample 2) after different durations of O_2 processing: (**a**) 5 s; (**b**) 20 s. Scale bar: 1 μm.

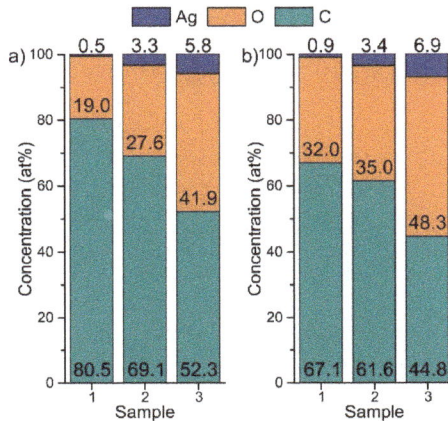

Figure 5. Chemical composition of DLC:Ag films (Samples 1–3) after 5 s (**a**) and 20 s (**b**) of O_2 plasma processing.

From Figure 5, one can see that after O_2 plasma processing, the surface concentration of carbon decreases and the amount of silver increases. Analysis of the composition of the films has shown a relatively high concentration of oxygen in the films as deposited and exposed to oxygen plasma. Unfortunately, these measurements were done after a long exposure of the samples in atmosphere, and potential changes of oxygen concentration in the films were hindered by the surface adsorption processes. Therefore, these values probably could be considered only to identify possible trends of variations but not the absolute (relative) concentration of oxygen.

2.3. Antimicrobial Activity of Bandage Prototype

Based on the best antimicrobial activity results, the 20 s plasma-processed DLC:Ag nanocomposite film containing 3.12 at % Ag was selected for further investigation as a building block of the bandage prototype. Synthetic silk substrate was used as a DLC:Ag substrate. A protective layer from cellulose sheet (0.01 mm) gelatin (0.1–0.15 mm) worked as a silver ion accumulation matrix and prevented the bandage from sticking to the wound's soft tissues. The antimicrobial properties of the prepared prototype were tested using the spread plate technique.

ASS experiments were used to follow the silver ion extraction process from the prototype containing the DLC:Ag nanocomposite layer. Figure 6 depicts silver ion concentration changes for different soaking durations of Sample 2 in purified water.

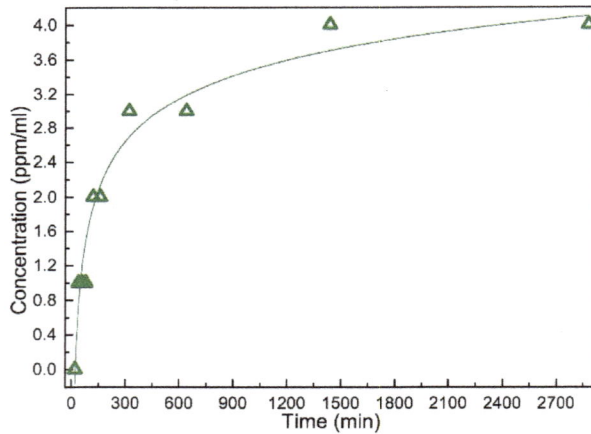

Figure 6. Silver ion concentration in purified water after soaking of synthetic silk coated with DLC:Ag (Sample 2 (3.12 at % Ag)) for different time durations obtained with AAS.

It was obtained that the silver ion concentration increases sharply during the first 300 min and saturates approximately after 900 min. A double logarithm function ($y = a \times \ln(-b \times \ln(x))$) can be used to approximate the experimental curve, and a high correlation coefficient $R^2 = 0.96$ was obtained for the constants a = 4.36 and b = −0.32.

Antimicrobial testing results of the smart bandage without and with the protective layer obtained using the spread-plate technique with four bacteria strains are summarized in Figure 7.

It was found that the exponential law (equation: $y = y_0 + A \times \exp(R_0 \times x)$) could be used to describe the time dependencies of CFU *versus* time, and Table 2 summarizes the values of the coefficients used in the approximations.

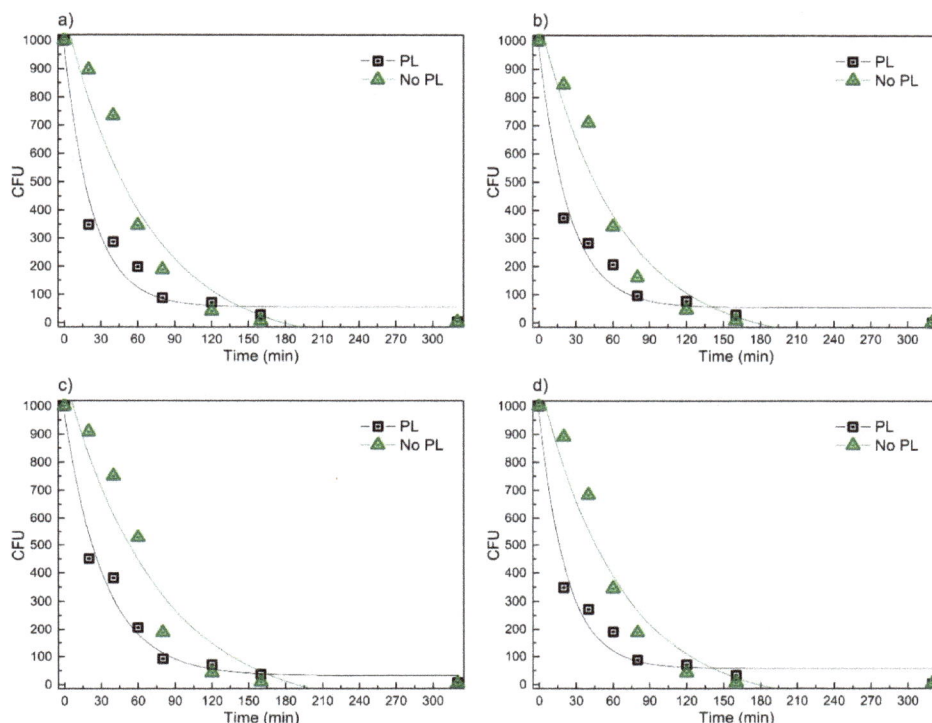

Figure 7. Time dependencies of bacteria colony forming units (CFU) *versus* time using the spread-plate technique for the bandage prototype (PL) and the reference sample (No PL) measured with four types of *S. aureus* bacteria strains: (**a**) LTSaDA01; (**b**) LTSaM01; (**c**) LTSa635 and (**d**) LTSa603.

Table 2. Coefficients of exponential dependencies used in the approximations of CFU *versus* time for four types of *S. aureus* bacteria strains.

Sample Structure	Fitting Coefficients	LTSaDA01 (a)		LTSaM01 (b)		LTSa635 (c)		LTSa603 (d)	
		Value	S.E. [1]	Value	S.E.	Value	S.E.	Value	S.E.
PL	y_0	55.37	36.36	54.13	34.42	30.77	35.16	56.566	33.60
	A	921.2	74.29	922.4	68.84	943.4	61.89	922.87	69.53
	R_0	−0.0430	0.00826	−0.0407	0.00717	−0.0305	0.00463	−0.0445	0.00803
	R^2	0.956		0.962		0.971		0.961	
No PL	y_0	−83.45	110.3	−75.39	97.95	−101.1	130.6	−77.21	95.29
	A	1184.3	134.41	1157.7	121.45	1208.68	149.92	1170.6	117.74
	R_0	−0.0149	0.00408	−0.0154	0.0039	−0.0132	0.00397	−0.0153	0.00371
	R^2	0.919		0.930		0.908		0.936	

[1] Standard Error.

Tests with bacteria LTSaDA01, LTSaM01 and LTSa603 revealed that after 20 min, DLC:Ag-coated synthetic silk without PL killed 10%–15% of *S. aureus* bacteria, when the same material with PL killed 63%–65%. Tests with LTSa635 (MRSA) for the same duration of time have shown efficiencies of 9% and 55%, respectively. Results of the longest soaking duration for 320 min revealed that the bandage without PL had the same antimicrobial effect as compared to the bandage plated with PL, *i.e.*, after bandages, in both cases, 99% of all bacteria were killed, and only a few CFU were observed.

3. Discussion

As was presented above, DC-reactive magnetron sputtering appears as an efficient way to produce nanocomposite DLC:Ag coatings. Simple variation of the Ar/C_2H_2 flux ratio enables the deposition of films with variable content of silver nanoparticle filler embedded in DLC matrix. The typical structure of the DLC:Ag films deposited under similar conditions can be found in [27]. Similar results were reported, as well, by [28], where DC-reactive unbalanced magnetron sputtering of Ag target in acetylene atmosphere allowed producing DLC:Ag nanocomposite thin films with a variable content of silver. During the experiments, the thickness of the film deposited on crystalline silicon and textile was approximately 40 nm, and according to Figure 1b,c, the average size of silver nanoparticles varied within a range 2–63 nm. In the case of lower silver concentration, low dimension silver nanoparticles prevailed. The average diameters of silver nanoclusters in DLC:Ag nanocomposite films were 23.7 nm and 28.8 nm for Samples 2 and 3, respectively. The larger silver content in the coatings matrix correlated with larger silver cluster diameter. This is in good agreement with our previous experiments, where we have performed deposition of DLC:Ag nanocomposites on silica substrates [29,30].

Antimicrobial results demonstrated that as-deposited DLC:Ag samples with low silver concentration (Samples 1–3) had no or only very weak antimicrobial effect (see Figure 2). As we have defined, additional O_2 plasma etching appears as an efficient tool in enhancing the antibacterial effect of the DLC:Ag nanocomposite surface. According to the results presented in Figure 5, O_2 plasma processing reduces the carbon surface concentration and provides the developed nanocomposite surface (Figure 4), which seems close to the results obtained in [31], where O_2 etching of organic materials provided nanotextured surfaces. It should be noted that plasma etching for 20 s reduced carbon content by 13.9% on average. During this process, a thin layer of carbon from the DLC:Ag surface was removed to expose more silver nanoparticles, which were embedded and covered by the DLC matrix [32]. On the other hand, due to O_2 plasma bombardment, coalescence of silver particles [33] and partial oxidation of silver nanoparticles take place. From this point of view, Sample 2 appears to be more efficient as compared to Sample 3 in terms of antimicrobial properties due to the smaller size of the nanoparticles. The dominating mechanisms could be elucidated after more comprehensive analysis of the behavior of DLC:Ag, and these experiments are in progress. After the oxygen plasma etching procedure, the average silver content increased by 0.5 at % after 20 s. The reduction of Ag in DLC:Ag films in the case of the highest silver concentration (Sample 3) could be explained by volume expansion and strain-induced cracking of oxidized Ag NPs due to oxygen ion bombardment [34,35]. Probably, a high concentration of silver, as well as an increase of the size of the nanoparticles in the case of Sample 3 (as compared to the other two samples) contribute to the efficiency of the mentioned mechanism. According to [36], the improvement of antimicrobial properties correlates with the increased hydrophobicity and, according to our findings, with the increased surface concentration of silver, as well.

As one can see, the etching time affected structural surface changes (Figure 4). Comparing samples etched for 5 s and as-deposited samples, only minor surface changes were observed, but the samples exposed to O_2 plasma for the longer time had many small dimples instead of a smooth surface. According to [36], this surface process is defined by the reaction of carbon materials, like DLC films with oxygen plasma, where plasma produces a destruction of the graphite rings. This results in an increase of single carbon chains where the concentration of aliphatic carbons atoms is pre-dominant in O_2 plasma-treated DLC films. Oxygen plasma treatment makes the DLC surface more desorbed, rougher and superhydrophilic [36]. Finally, the superhydrophilic surface ensured maximum silver ion diffusion from the samples to wet agar media. On the other hand, we have found that the prolonged plasma exposure (above 25 s) resulted in the decrease of antibacterial activity for all investigated strains and investigated silver concentrations (see Figure 3). This effect could be attributed to the ion beam irradiation-induced increase of silver nanoparticles (ripening process) that we have observed early in the case of reactive ion etching of DLC:Ag nanocomposites [33]. One can expect that the larger particles have a smaller surface area, which is responsible for silver ion release into media. The nanoparticle size effect on the antimicrobial properties was reported as well in [37,38], where the authors declare better antibacterial properties of small dimensions of AgNPs.

According to our findings, the optimal time for RF oxygen plasma etching was 20–25 s (see Figure 3). For such plasma exposure, all of the coatings with different silver content revealed the best antimicrobial properties, *i.e.,* they had the developed surface and optimum dimensions of the silver nanoparticles.

For such kinds of samples (Sample 2 as a typical example was taken (silver content after 20 s; etching was 3.41 at %)), AAS analysis revealed efficient silver ion migration into aqueous media (Figure 6). We found out that saturation of silver ion concentration in purified water takes place, and after 24 h, at a 35 °C temperature, it reaches 4 ppm. It should be noted that according to [38], the antibacterial activity against *S. aureus* starts at about 1 ppm. These results correlate well with the microbiological testing (spread-plate technique; Figure 7) data where larger silver concentrations killed more bacteria, as well.

Experimental data (Figure 7) revealed that the bandage without PL (soaking duration of 20 min) demonstrated a lower antimicrobial effect as compared to the bandage where PL was applied. One can assume that our chosen aqueous media (PL) can accumulate silver ions inside. After 24 h of exposure of the bandage, the PL has accumulated silver ions and during the test acted as an efficient source of silver ions. Moreover, the agar-gelatin layer at 35 °C can dissolve easily, and silver ions can spread rapidly to all aqueous media, providing a very good antimicrobial effect. In all studied cases, PL technology improved the antibacterial properties of the bandage, and it was more than approximately 50% effective during the same time interval. The action speed was the main advantage of the prototype, *i.e.,* PL can be used as silver ion accumulation media for fast ion release in aqueous media, enabling to kill instantly more than 50% of all of the bacteria population. It should be noted that in wound healing, this is a very important factor, as the bacteria need to be killed in a short period of time to prevent efficient growth of the bacteria population [39–41]; e.g., in [16], it was shown that bacterial adhesion involved reversible bacterial association in the first 1–2 h after post-implantation, followed by stronger bacterial adhesion, approximately 2–3 h later. After 24 h, certain bacteria formed a biofilm, which was resistant to host defense and systemic antibiotic treatment.

In addition to the efficient antimicrobial properties, our bandage with PL having a gel-like structure with the synthetic silk skeleton ensures good mechanical properties, as well. The cellulose sheet as the membrane can sustain small particles and debris inside the bandage close to the DLC:Ag surface, avoiding wound contamination. The gelatin and agar layer has very good water absorption abilities, as was demonstrated in [42]. Furthermore, according to [14], approximately 39% more silver is released into alkaline sweat (pH 8.0) as compared to acidic sweat (pH 5.5). In infected wound, pH usually moves to a neutral or slightly alkali pH value; in that case, our technology also has an advantage: silver ions should migrate faster, from nanoparticles inside the DLC matrix, and the antimicrobial effect could be even stronger comparing to the tests in saline solution (0.9%) [14]. The higher release rate of Ag ion concentration into the surrounding medium and the longer it is sustained, the more thorough the antimicrobial effect will be [43].

It should be noted as well that cytotoxicity is one of the main problems with using nanotechnology in medical devices like bandages. Embedded cells like nanoparticles can cause adverse side effects to the organism. In our bandage prototype, only the PL structure was used, to avoid this problem, but further tests are needed to investigate toxicity. In the prototype bandage, the silver ion concentration, according to the AAS data, could reach 4 ppm or 4 µg/mL. According to [13], the minimum inhibitory concentrations of all bacteria tested were in a concentration range of AgNPs (between 3.37 and 13.5 µg/mL) in which there was no observed significant cytotoxic activity compared to the control [13].

4. Materials and Methods

4.1. Deposition, Characterization and O₂ Plasma Processing of DLC:Ag Films

DLC:Ag coatings with different Ag content were deposited on textile (twill weaved synthetic silk, with a weft density of 110 cm^{-1} and a warp density of 100 cm^{-1}) by DC-reactive unbalanced magnetron sputtering of Ag target in acetylene atmosphere. Ag content was controlled varying feed stock gas flow rates and changing magnetron power, as well as bias voltage. Crystalline silicon wafer substrates were used as well for the comparative control of the deposition process. The deposition conditions are summarized in Table 3. Further details on the deposition and properties of DLC:Ag coatings with variable silver concentration can be found in our early papers [33,44,45].

Table 3. Deposition conditions of DLC:Ag films.

Sample No.	Sputtering Duration (s)	Ar Gas Flow (sccm)	C₂H₂ Gas Flow (sccm)	Magnetron Voltage (V)	Magnetron Current (A)
1	520	70	21.1	553–625	0.07–0.12
2	235	70	21.1	568–741	0.07–0.22
3	200	80	7.8	625–656	0.10–0.11

The thickness of the deposited DLC:Ag coatings was approximately 40 nm (as measured by scanning electron microscope (SEM)). The AgNPs' particle size distribution and chemical composition were obtained employing SEM FEI Quanta 200 FEG with an energy dispersive X-ray spectrometer (EDS), Bruker Quantax. NPs' diameters and the chemical composition of the films were studied for the DLC:Ag samples deposited on Si substrates. EDS measurements were performed at a5-keV accelerating voltage in order to minimize the excitation of the Si K$_\alpha$ peak. NP size analysis was preformed employing ImageJ (NIH) software and custom MATLAB (MathWorks) code. More detailed information about the applied NP analysis procedures can be found elsewhere [33,46]. Measurements of as-deposited nanocomposite thin film were performed at 3–5 different points, and average normalized surface concentrations (excluding silicon) can be found in Table 1.

After the deposition, the samples were additionally etched by radio frequency (RF; 13.56 MHz) oxygen plasma (99.9%) in 133 Pa pressure and 0.3 W/cm² power for 5–30 s. Morphology and antimicrobial properties of the as-deposited and O₂ plasma-etched DLC:Ag films on silicon and silk were investigated.

4.2. Microbiological Testing of Virgin and O₂ Plasma-Processed DLC:Ag Films

For testing of the bactericidal activity of DLC:Ag films on multiplying bacteria, four clinically-important strains causing skin and wound infections of *S. aureus* previously isolated at Lithuanian University of Health Sciences were selected. Two strains were isolated from sick humans (LTSaDA01 and LTSaM01), as well as two strains were isolated from diseased pet animals: a dog (LTSa603) and a cat (LTSa635). The strain LTSa635 was methicillin-resistant. The antimicrobial properties of the DLC:Ag layer, as well as bandage prototype including the DLC:Ag layer were tested using the disk diffusion method (see Figure 8a) and the spread-plate technique [47] (see Figure 8b). The bacterial suspension density of 1 McFarland unit in saline solution (0.9%) was prepared and inoculated onto Mueller Hinton agar (Thermo Scientific, Leicestershire, UK) in 94 mm-diameter Petri dishes. In the disk diffusion method, synthetic silk samples (virgin and etched) of a 6 mm × 6 mm (±1 mm) size with DLC:Ag coatings (Samples 1–3) were glued onto the agar surface using a small drop of water. The samples were incubated for 24 h at 35 °C. The dimensions of inhibition zones were measured from four sides of the sample, and the average value of the clear zone was calculated. To check the temporal stability of the coatings, experiments with all three different silver concentration DLC:Ag samples on synthetic silk were repeated for 4 times during a one-month period: at first day and thereafter at the 1st, 2nd and 3rd week after the deposition. Experimental results were fitted employing OriginPro (OriginLab, Northampton, MA, USA).

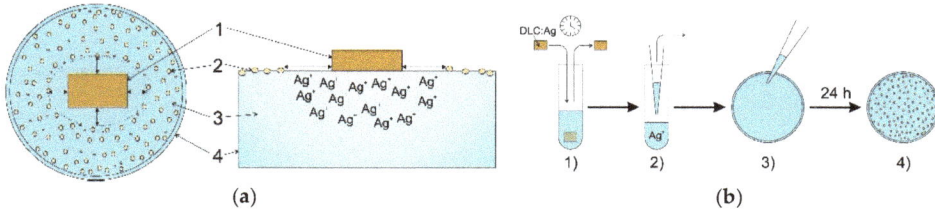

(a)

(b)

Figure 8. Antimicrobial testing techniques. (**a**) Schematics of the disk diffusion method used for microbiological testing (Ag$^+$ diffusion of silver ions from the DLC:Ag surface into agar): (**1**) sample coated with DLC:Ag nanocomposite; (**2**) bacteria colonies; (**3**) agar (bacteria nutrition media); (**4**) Petri dish; (**b**) Schematics of the spread plate technique: (**1**) bandage prototype soaked in a test tube with bacteria saline solution for different time durations; (**2**) the bacterial solution with Ag$^+$ ions (0.1 mL) was transported into a Petri dish; (**3**) inoculation of bacteria onto the agar surface; (**4**) calculation of colony forming units (CFU).

4.3. Construction of the Bandage Prototype

DLC:Ag coating on synthetic silk indicated the best antimicrobial properties and was used as a part of the smart bandage. The coating was exposed to UV irradiation for disinfection. As a protective layer (PL), cellulose fibers (medium, C6288 Sigma, Sigma Aldrich, St. Louis, MO, USA), gelatin (53028 FLUKA, Sigma Aldrich) and agar (A1296 SIGMA, Sigma Aldrich) were used. A thin cellulose sheet, acting as a membrane, was manufactured from microfiber cellulose and gelatin as a gluing material. It was rolled into a 0.01 mm-thick sheet. The cellulose sheet was glued using gelatin and pressed using a metal roller on DLC:Ag-coated synthetic silk. A hot suspension from gelatin and agar (90%/10%, respectively) was prepared and placed on top as a second layer using a spin coater at low speed (120 rpm). The structure of the smart bandage is presented in Figure 9. In such way, the prepared bandage prototype was used in the antibacterial tests after 24 h.

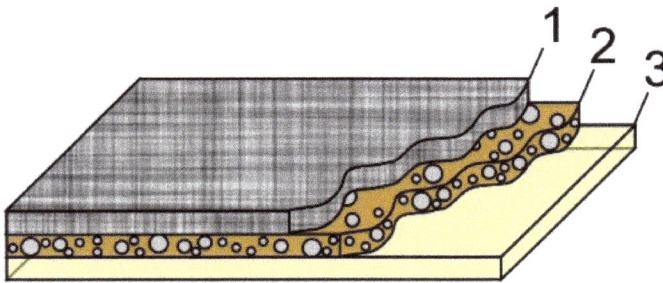

Figure 9. Principle structure of the proposed bandage: (**1**) synthetic silk substrate; (**2**) thin nanocomposite DLC:Ag films etched for 20 s in O$_2$ plasma; (**3**) protective layer made from the thin cellulose sheet (membrane) and the gelatin layer.

4.4. Antimicrobial Testing of the Bandage Prototype

Bactericidal activity evaluation tests of the bandage prototype were carried out using the same four strains of *S. aureus* as described Section 4.2. The spread-plate technique [47] was applied to confirm the bandage's ability to kill bacteria on the surface and in the liquid media around it. The protective layer-covered rectangular-shaped bandage of 100 mm² in area was soaked for 20, 40, 60, 80, 120, 160, 320 min in a thermostat at 35 °C in a 1-mL volume of bacteria in saline solution of 0.1 McFarland units. To check the efficiency of the protective layer, an identical experiment was carried out with the bandage including just the nanocomposite layer (Sample 2 etched for 20 s). After incubation, the withdrawn solution was diluted 1:1000 times, and thereafter, 100 μL of bacteria suspension were inoculated onto a Petri dish containing Mueller Hinton. After 24 h, colony forming units (CFU) were counted. For statistical reliability, this experiment was repeated 3 times.

4.5. Antimicrobial Testing of the Bandage Prototype

Atomic absorption spectroscopy (AAS) was used to identify Ag^+ migration from the bandage (synthetic silk with Ag nanoparticles) to ultrapure water. The disk-shaped 10-cm² bandage was soaked in the test tube with 10 mL of thermostated water at 35 °C for time intervals from 20 min to 48 h. Later on, water was filtrated from large (0.1 mm) particles, and AAS measurements were performed. A Perkin Elmer Model 403 spectrometer was employed.

5. Conclusions

Diamond-like carbon-based silver nanocomposite layers deposited by DC-reactive magnetron sputtering on textile appeared as an effective source of Ag^+ ions and demonstrated expressed antibacterial properties against four tested strains of *Staphylococcus aureus* bacteria.

The efficiency of the Ag^+ ion release to the aqueous media can be increased by further RF oxygen plasma etching of the nanocomposite. It was obtained that the best antibacterial properties were demonstrated by the plasma-processed DLC:Ag layer having a 3.12 at % Ag surface concentration with the dominating linear dimensions of nanoparticles being 23.7 nm.

An extra protective layer made from cellulose and gelatin with agar contributed to the accumulation and efficient release of silver ions to the aqueous media, increasing the bandage antimicrobial efficiency up to 50% as compared to the single DLC:Ag layer on textile.

The proposed bandage prototype (having a silver ion concentration in the protective layer below the toxic level for organism cells) was able to kill more than 99.9% of all strains of bacteria after 320 min, including methicillin-resistant *Staphylococcus aureus*.

Acknowledgments: This research was funded by the common Lithuanian University of Health Sciences and Kaunas University of Technology Research Project "NANOSMARTPLASTER" Grant No. PP38/152. No funds for covering the costs to publish in open access were received. Authors acknowledge Igoris Prosyčevas and Irina Abelit for atomic absorption spectroscopy measurements.

Author Contributions: S.T. and T.T. conceived of and designed the experiments. T.J., M.R., R.Š., I.J. and A.V. performed the experiments. A.J., T.J. and T.T. analyzed the data. T.T. and M.R. contributed reagents/materials/ analysis tools. T.J. and T.T. wrote the paper.

Conflicts of Interest: The authors declare no conflict of interest.

Abbreviations

The following abbreviations are used in this manuscript

DLC	diamond-like carbon
DLC:Ag	diamond-like carbon with silver nanoparticles
RF	radio frequency
MRSA	methicillin-resistant *Staphylococcus aureus*
S. aureus	*Staphylococcus aureus*
Ag^+	silver ions
DNA	deoxyribonucleic acid
Ag NPs	silver nanoparticles
DC	direct current
PL	protective layer

SEM	scanning electron microscopy
EDS	energy dispersive X-ray spectroscopy
AAS	atomic absorption spectroscopy
N	number of analyzed particles
d_{av}	average particle diameter
a	constant
b	constant
R^2	correlation coefficient
y_0	coefficient of extrapolation
A	coefficient of extrapolation
R_0	coefficient of extrapolation
S.E.	standard error

References

1. Nanda, A.; Saravanan, M. Biosynthesis of silver nanoparticles from staphylococcus aureus and its antimicrobial activity against MRSA and MRSE. *Nanomed. Nanotechnol. Biol. Med.* **2009**, *5*, 452–456. [CrossRef] [PubMed]
2. Thomas, S. MRSA and the Use of Silver Dressings: Overcoming Bacterial Resistance. Available online: http://www.worldwidewounds.com/2004/november/Thomas/Introducing-Silver-Dressings.html (accessed on 30 March 2016).
3. Silver, L.L. Multi-targeting by monotherapeutic antibacterials. *Nat. Rev. Drug Discov* **2007**, *6*, 41–55. [CrossRef] [PubMed]
4. Schwarz, F.P.; Hauser-Gerspach, I.; Waltimo, T.; Stritzker, B. Antibacterial properties of silver containing diamond like carbon coatings produced by ion induced polymer densification. *Surf. Coat. Technol.* **2011**, *205*, 4850–4854. [CrossRef]
5. Thet, N.T.; Alves, D.R.; Bean, J.E.; Booth, S.; Nzakizwanayo, J.; Young, A.E.R.; Jones, B.V; Jenkins, A.T.A. Prototype development of the intelligent hydrogel wound dressing and its efficacy in the detection of model pathogenic wound biofilms. *ACS Appl. Mater. Interfaces* **2015**. [CrossRef] [PubMed]
6. Bociaga, D.; Komorowski, P.; Batory, D.; Szymanski, W.; Olejnik, A.; Jastrzebski, K.; Jakubowski, W. Silver-doped nanocomposite carbon coatings (Ag-DLC) for biomedical applications-physiochemical and biological evaluation. *Appl. Surf. Sci.* **2015**, *355*, 388–397. [CrossRef]
7. David, M.Z.; Daum, R.S. Community-associated methicillin-resistant staphylococcus aureus: Epidemiology and clinical consequences of an emerging epidemic. *Clin. Microbiol. Rev.* **2010**, *23*, 616–687. [CrossRef] [PubMed]
8. Hiramatsu, K.; Katayama, Y.; Matsuo, M.; Sasaki, T.; Morimoto, Y.; Sekiguchi, A.; Baba, T. Multi-drug-resistant staphylococcus aureus and future chemotherapy. *J. Infect. Chemother.* **2014**, *20*, 593–601. [CrossRef] [PubMed]
9. Paladini, F.; Picca, R.A.; Sportelli, M.C.; Cioffi, N.; Sannino, A.; Pollini, M. Surface chemical and biological characterization of flax fabrics modified with silver nanoparticles for biomedical applications. *Mater. Sci. Eng. C Mater. Biol. Appl.* **2015**, *52*, 1–10. [CrossRef] [PubMed]
10. Owens, B. Silver makes antibiotics thousands of times more effective. *Nature News*, 19 June 2013. [CrossRef]
11. Knetsch, M.L.W.; Koole, L.H. New strategies in the development of antimicrobial coatings: The example of increasing usage of silver and silver nanoparticles. *Polymers* **2011**, *3*, 340–366. [CrossRef]
12. Koerner, E.; Aguirre, M.H.; Fortunato, G.; Ritter, A.; Ruehe, J.; Hegemann, D. Formation and distribution of silver nanoparticles in a functional plasma polymer matrix and related Ag+ release properties. *Plasma Processes Polym.* **2010**, *7*, 619–625. [CrossRef]
13. Quelemes, P.V.; Araruna, F.B.; de Faria, B.E.F.; Kuckelhaus, S.A.S.; da Silva, D.A.; Mendonca, R.Z.; Eiras, C.; Soares, M.J.d.S.; Leite, J.R.S.A. Development and antibacterial activity of cashew gum-based silver nanoparticles. *Int. J. Mol. Sci.* **2013**, *14*, 4969–4981. [CrossRef] [PubMed]
14. Lazic, V.; Saponjic, Z.; Vodnik, V.; Dimitrijevic, S.; Jovancic, P.; Nedeljkovic, J.; Radetic, M. A study of the antibacterial activity and stability of dyed cotton fabrics modified with different forms of silver. *J. Serbian Chem. Soc.* **2012**, *77*, 225–234. [CrossRef]
15. Palomba, M.; Carotenuto, G.; Cristino, L.; Di Grazia, M.A.; Nicolais, F.; De Nicola, S. Activity of antimicrobial silver polystyrene nanocomposites. *J. Nanomater.* **2012**. [CrossRef]

16. Jamuna-Thevi, K.; Bakar, S.A.; Ibrahim, S.; Shahab, N.; Toff, M.R.M. Quantification of silver ion release, *in vitro* cytotoxicity and antibacterial properties of nanostuctured Ag doped TiO$_2$ coatings on stainless steel deposited by rf magnetron sputtering. *Vacuum* **2011**, *86*, 235–241. [CrossRef]

17. Alissawi, N.; Zaporojtchenko, V.; Strunskus, T.; Hrkac, T.; Kocabas, I.; Erkartal, B.; Chakravadhanula, V.S.K.; Kienle, L.; Grundmeier, G.; Garbe-Schoenberg, D.; *et al.* Tuning of the ion release properties of silver nanoparticles buried under a hydrophobic polymer barrier. *J. Nanopart. Res.* **2012**, *14*. [CrossRef]

18. Larese, F.F.; D'Agostin, F.; Crosera, M.; Adami, G.; Renzi, N.; Bovenzi, M.; Maina, G. Human skin penetration of silver nanoparticles through intact and damaged skin. *Toxicology* **2009**, *255*, 33–37. [CrossRef] [PubMed]

19. Tolaymat, T.M.; El Badawy, A.M.; Genaidy, A.; Scheckel, K.G.; Luxton, T.P.; Suidan, M. An evidence-based environmental perspective of manufactured silver nanoparticle in syntheses and applications: A systematic review and critical appraisal of peer-reviewed scientific papers. *Sci. Total Environ.* **2010**, *408*, 999–1006. [CrossRef] [PubMed]

20. Zhou, H.; Xu, L.; Ogino, A.; Nagatsu, M. Investigation into the antibacterial property of carbon films. *Diam. Relat. Mater.* **2008**, *17*, 1416–1419. [CrossRef]

21. Silver as an Anti-Bacterial. Available online: https://www.silverinstitute.org/site/silver-in-technology/silver-in-medicine/bandages/ (accessed on 30 March 2016).

22. Elastoplast-Antibacterial Waterproof. Available online: http://www.en.elastoplast.ca/Products/waterproof-antibacterial (accessed on 30 March 2016).

23. Band-aid Plus Antibiotic Adhesive Bandages, Assorted Sizes. Available online: http://www.drugstore.com/band-aid-plus-antibiotic-adhesive-bandages-assorted-sizes/qxp149702 (accessed on 30 March 2016).

24. Percival, S.L.; Bowler, P.G.; Dolman, J. Antimicrobial activity of silver-containing dressings on wound microorganisms using an *in vitro* biofilm model. *Int. Wound J.* **2007**, *4*, 186–191. [CrossRef] [PubMed]

25. Newman, G.R.; Walker, M.; Hobot, J.A.; Bowler, P.G. Visualisation of bacterial sequestration and bactericidal activity within hydrating hydrofiber® wound dressings. *Biomaterials* **2006**, *27*, 1129–1139. [CrossRef] [PubMed]

26. Wach, R.A.; Mitomo, H.; Nagasawa, N.; Yoshii, F. Radiation crosslinking of carboxymethylcellulose of various degree of substitution at high concentration in aqueous solutions of natural pH. *Radiat. Phys. Chem.* **2003**, *68*, 771–779. [CrossRef]

27. Meskinis, S.; Vasiliauskas, A.; Slapikas, K.; Gudaitis, R.; Andrulevicius, M.; Ciegis, A.; Niaura, G.; Kondrotas, R.; Tamulevicius, S. Bias effects on structure and piezoresistive properties of DLC:Ag thin films. *Surf. Coat. Technol.* **2014**, *255*, 84–89. [CrossRef]

28. Choi, H.W.; Dauskardt, R.H.; Lee, S.-C.; Lee, K.-R.; Oh, K.H. Characteristic of silver doped DLC films on surface properties and protein adsorption. *Diam. Relat. Mater.* **2008**, *17*, 252–257. [CrossRef]

29. Yaremchuk, I.; Meskinis, S.; Fitio, V.; Bobitski, Y.; Slapikas, K.; Ciegis, A.; Balevicius, Z.; Selskis, A.; Tamulevicius, S. Spectroellipsometric characterization and modeling of plasmonic diamond-like carbon nanocomposite films with embedded ag nanoparticles. *Nanoscale Res. Lett.* **2015**, *10*. [CrossRef] [PubMed]

30. Meškinis, Š.; Vasiliauskas, A.; Šlapikas, K.; Gudaitis, R.; Yaremchuk, I.; Fitio, V.; Bobitski, Y.; Tamulevičius, S. Annealing effects on structure and optical properties of diamond-like carbon films containing silver. *Nanoscale Res. Lett.* **2016**, *11*, 1–9. [CrossRef] [PubMed]

31. Tsougeni, K.; Vourdas, N.; Tserepi, A.; Gogolides, E.; Cardinaud, C. Mechanisms of oxygen plasma nanotexturing of organic polymer surfaces: From stable super hydrophilic to super hydrophobic surfaces. *Langmuir* **2009**, *25*, 11748–11759. [CrossRef] [PubMed]

32. Hesse, E.; Creighton, J.A. Investigation by surface-enhanced raman spectroscopy of the effect of oxygen and hydrogen plasmas on adsorbate-covered gold and silver island films. *Langmuir* **1999**, *15*, 3545–3550. [CrossRef]

33. Tamulevicius, T.; Tamuleviciene, A.; Virganavicius, D.; Vasiliauskas, A.; Kopustinskas, V.; Meskinis, S.; Tamulevicius, S. Structuring of DLC:Ag nanocomposite thin films employing plasma chemical etching and ion sputtering. *Nucl. Instrum. Methods Phys. Res. Sect. B-Beam Interact. Mater. Atoms* **2014**, *341*, 1–6. [CrossRef]

34. Zeng, Y.X.; Chen, L.H.; Zou, Y.L.; Nguyen, P.A.; Hansen, J.D.; Alford, T.L. Processing and encapsulation of silver patterns by using reactive ion etch and ammonia anneal. *Mater. Chem. Phys.* **2000**, *66*, 77–82. [CrossRef]

35. Nguyen, P.; Zeng, Y.X.; Alford, T.L. Reactive ion etch of patterned and blanket silver thin films in Cl^{-2}/O^{-2} and O^{-2} glow discharges. *J. Vac. Sci. Technol. B* **1999**, *17*, 2204–2209. [CrossRef]

36. Marciano, F.R.; Bonetti, L.F.; Da-Silva, N.S.; Corat, E.J.; Trava-Airoldi, V.J. Wettability and antibacterial activity of modified diamond-like carbon films. *Appl. Surf. Sci.* **2009**, *255*, 8377–8382. [CrossRef]

37. Cao, H.; Qiao, Y.; Liu, X.; Lu, T.; Cui, T.; Meng, F.; Chu, P.K. Electron storage mediated dark antibacterial action of bound silver nanoparticles: Smaller is not always better. *Acta Biomater.* **2013**, *9*, 5100–5110. [CrossRef] [PubMed]

38. Liu, H.-L.; Dai, S.A.; Fu, K.-Y.; Hsu, S.-H. Antibacterial properties of silver nanoparticles in three different sizes and their nanocomposites with a new waterborne polyurethane. *Int. J. Nanomed.* **2010**, *5*, 1017–1028.

39. Ip, M.; Lui, S.L.; Poon, V.K.M.; Lung, I.; Burd, A. Antimicrobial activities of silver dressings: An *in vitro* comparison. *J. Med. Microbiol.* **2006**, *55*, 59–63. [CrossRef] [PubMed]

40. Wright, J.B.; Lam, K.; Burrell, R.E. Wound management in an era of increasing bacterial antibiotic resistance: A role for topical silver treatment. *Am. J. Infect. Control* **1998**, *26*, 572–577. [CrossRef] [PubMed]

41. Edwards, R.; Harding, K.G. Bacteria and wound healing. *Curr. Opin. Infect. Dis.* **2004**, *17*, 91–96. [CrossRef] [PubMed]

42. Vimala, K.; Yallapu, M.M.; Varaprasad, K.; Reddy, N.N.; Ravindra, S.; Naidu, N.S.; Raju, K.M. Fabrication of curcumin encapsulated chitosan-PVA silver nanocomposite films for improved antimicrobial activity. *J. Biomater. Nanobiotechnol.* **2011**, *2*, 55–64. [CrossRef]

43. Yilma, A.N.; Singh, S.R.; Dixit, S.; Dennis, V.A. Anti-inflammatory effects of silver-polyvinyl pyrrolidone (Ag-PVP) nanoparticles in mouse macrophages infected with live chlamydia trachomatis. *Int. J. Nanomed.* **2013**, *8*, 2421–2432.

44. Meskinis, S.; Vasiliauskas, A.; Slapikas, K.; Niaura, G.; Juskenas, R.; Andrulevicius, M.; Tamulevicius, S. Structure of the silver containing diamond like carbon films: Study by multiwavelength raman spectroscopy and xrd. *Diam. Relat. Mater.* **2013**, *40*, 32–37. [CrossRef]

45. Yaremchuk, I.; Tamuleviciene, A.; Tamulevicius, T.; Slapikas, K.; Balevicius, Z.; Tamulevicius, S. Modeling of the plasmonic properties of DLC-Ag nanocomposite films. *Phy. Status Solidi A-Appl. Mater. Sci.* **2014**, *211*, 329–335. [CrossRef]

46. **Meškinis, Š.; Tamulevičius, T.; Niaura, G.; Šlapikas, K.; Vasiliauskas, A.; Ulčinas, O.; Tamulevičius, S.** Surface enhanced raman scattering effect in diamond like carbon films containing ag nanoparticles. *J. Nanosci. Nanotechnol.* **2016**, *16*, 1–9.

47. Seil, J.T.; Webster, T.J. Antimicrobial applications of nanotechnology: Methods and literature. *Int. J. Nanomed.* **2012**, *7*, 2767–2781.

materials

MDPI

Article

Antimicrobial Silver Chloride Nanoparticles Stabilized with Chitosan Oligomer for the Healing of Burns

Yun Ok Kang [1], Ju-Young Jung [2], Donghwan Cho [3], Oh Hyeong Kwon [3], Ja Young Cheon [4] and Won Ho Park [4,*]

[1] Department of Nano Manufacturing Technology, Nano Convergence Mechanical Systems Research Division, Korea Institute of Machinery and Materials, Daejeon 34103, Korea; ok2733@gmail.com
[2] Department of Veterinary Medicine, Chungnam National University, Daejeon 305-764, Korea; jyjung@cnu.ac.kr
[3] Department of Polymer Science and Engineering, Kumoh National Institute of Technology, Gumi 39177, Korea; dcho@kumoh.ac.kr (D.C.); ohkwon@kumoh.ac.kr (O.H.K.)
[4] Department of Advanced Organic Materials and Textile System Engineering, Chungnam National University, Daejeon 305-764, Korea; alranim@nate.com
* Correspondence: parkwh@cnu.ac.kr; Tel.: +82-42-821-6613; Fax: +82-42-823-3736

Academic Editor: Mauro Pollini
Received: 11 February 2016; Accepted: 11 March 2016; Published: 23 March 2016

Abstract: Recently, numerous compounds have been studied in order to develop antibacterial agents, which can prevent colonized wounds from infection, and assist the wound healing. For this purpose, novel silver chloride nanoparticles stabilized with chitosan oligomer (CHI-AgCl NPs) were synthesized to investigate the influence of antibacterial chitosan oligomer (CHI) exerted by the silver chloride nanoparticles (AgCl NPs) on burn wound healing in a rat model. The CHI-AgCl NPs had a spherical morphology with a mean diameter of 42 ± 15 nm. The burn wound healing of CHI-AgCl NPs ointment was compared with untreated group, Vaseline ointment, and chitosan ointment group. The burn wound treated with CHI-AgCl NPs ointment was completely healed by 14 treatment days, and was similar to normal skin. Particularly, the regenerated collagen density became the highest in the CHI-AgCl NPs ointment group. The CHI-AgCl NPs ointment is considered a suitable healing agent for burn wounds, due to dual antibacterial activity of the AgCl NPs and CHI.

Keywords: Silver chloride nanoparticles; Chitosan oligomer; Burn wound healing; Antibacterial activity

1. Introduction

Antibacterial materials could be used in food packaging or handling, medical tools, hospitals where people are more vulnerable to infections, possible areas like bathrooms where personal hygiene is important, air and water filtration, *etc.* Among various materials known to be effective against a variety of bacteria, silver (Ag) compounds (including elemental silver, silver oxide, silver halide, *etc.*) are unique materials due to their powerful antibacterial activities against nearly 650 bacteria strains. In addition, Ag compounds are well-known and have already been used in many biological and medical fields, such as biosensors, wound healing materials, dental resin composites, and cancer therapeutics [1,2].

Traumatic wounds including burn wound occur frequently in skin loss, and bacterial infections often result from heavy contamination. There are many dressings and creams available to clinicians for use in treating burn wounds. Numerous agents are valued for their antibacterial activity, which can prevent colonized wounds from becoming infected, thereby assisting the wound to heal. Wound infection is one of the main problems and serious consequence in severe burns or extensive skin loss since bacteria produce leukocidin and tissue-destroying enzymes that further impair healing. As a result, wound infections alter and delay normal the wound healing mechanism. One of the approaches for treating a wound infection is the use of wound dressings containing antibacterial materials with a broad-spectrum of activity. Ag compounds have been widely found in wound dressings in various forms, such as elemental Ag (Ag metal, and nanocrystalline Ag), inorganic compounds (silver oxide, silver phosphate, silver chloride, silver sulfate, silver calcium-sodium phosphate, silver zirconium compound, and silver sulfadiazine), and organic complexes (silver zinc allantoinate, silver alginate, and silver carboxymethylcellulose) [3–5]. Recent evidence suggests that the antibacterial mechanism is due to the release of Ag^+ or Ag^0. In addition, this antibacterial effect accelerates wound healing [6,7]. The slow release of Ag ions is required for a continual bactericidal concentration of Ag ions in the wound. Silver chloride (AgCl) is colorless and its low solubility product ensures a long life and a slow release of silver ions for antibacterial property [8,9]. Indeed, there have been some reports on using AgCl for antibacterial agent in wound dressing [10–12].

Chitosan, a β-1,4-linked polysaccharide of glucosamine (2-amino-2-deoxy-β-D-glucose) with lesser amounts of N-acetylglucosamine, is a natural non-toxic biopolymer derived by deacetylation of chitin [13]. Chitosan oligomer (chitooligosaccharide, CHI) is easy to prepare by the acidic or enzymatic partial hydrolysis of chitosan. It has been reported that lower oligomers of chitosan are water-soluble and biologically active, through their solubility and activities are dependent on average degree of polymerization (DP) and the degree of deacetylation (DD) [14,15]. Chitosan and chitosan oligomer have pronounced antibacterial effects because of the presence of amino groups [16,17] Many studies have also been carried out on the use of chitosan and its derivatives as a wound healing accelerator by enhancing the functions of inflammatory cells, and there is good evidence that chitosan can beneficially influence every separate stage of wound healing [18–22]. Chitosan and chitosan oligomer are currently being explored as novel tools for wound and burn dressings, because of their immunostimulating, hemostatic, antibacterial, nontoxic, biocompatible, and biodegradable properties [23].

In previous study [24], AgCl NPs stabilized with CHI were prepared by green synthesis. The effect of the CHI, which was used as a resource of Cl ions and a stabilizing agent, on the formation reaction of the AgCl NPs was investigated. It was confirmed that the Cl ions remained around ammonium group in CHI molecule since the CHI was made by acidic hydrolysis using hydrochloric acid. Ag ions readily reacted with the Cl ions during the formation of AgCl NPs. The synergistic effect of CHI-AgCl NPs on the antibacterial activity was confirmed to be because it was prepared by the combination of CHI and AgCl NPs, known antibacterial materials. However, further studies on accelerating effect of the *in vivo* wound healing by antibacterial CHI-AgCl NPs have not been studied.

The present study was carried out to give an example of an effective burn wound healing agent based on synergistic effect from using both AgCl and CHI. We synthesized CHI-AgCl NPs using water-soluble chitosan oligomer by an environment-friendly method. Furthermore, the CHI-AgCl NPs ointment was prepared to evaluate the healing effect for burn wound using rat model with the expectation of applications in pharmaceutical and biomedical fields.

2. Materials and Methods

2.1. Preparation of CHI-AgCl NPs

Oligomeric chitosan (CHI, DD = 87%) was supplied by Hyosung Co. (Korea), and its composition is as follows: dimer 2.31, trimer 12.53, tetramer 15.11, pentamer 13.59, hexamer 8.86, heptamer 6.46, octamer 8.87, and nonamer or higher 32.27 mol %. Cl ions (Cl^-) remained in CHI molecule because CHI was made by acidic hydrolysis using chitosan in hydrochloric acid. CHI was used as a resource of Cl ions and stabilizing agent for preparing AgCl NPs. Distilled water was used as solvent in all the syntheses to provide benign environmental conditions in this system. In a typical preparation, 0.5 mL of 0.1 M $AgNO_3$ solution was added to 60 mL of 5% (w/v) CHI solution. After complete dissolution, the mixture was reacted in a three-necked glass-stopper flask fitted with a double-walled spiral condenser to prevent evaporation and heated to 70 °C for 300 min. All solution components were purged with nitrogen gas to eliminate oxygen. After the formation of CHI-AgCl NPs, the suspension was freeze-dried immediately at –85 °C to produce the powder CHI-AgCl NPs. A series of experiments were performed to obtain transparent CHI-AgCl NPs by varying the order of reactants and the reaction temperature.

2.2. Characterization of CHI-AgCl NPs

A UV-Vis spectrophotometer (Shimadzu, UV-2450, Tokyo, Japan) was used to record absorption spectra in the suspension of CHI-AgCl NPs using a cell with path length of 1.0 cm. The morphology of CHI-AgCl NPs was observed using a transmission electron microscope (TEM) (EM 912 OMEGA, ZEISS, Jena, Germany). The samples for TEM observation were prepared by spotting a few microliters of the suspension of synthesized CHI-AgCl NPs onto a holey carbon TEM grid followed by drying before putting them into the TEM sample chamber.

2.3. Preparation of Ointments

Three ointments for *in-vivo* burn wound healing were prepared from Vaseline, CHI, and CHI-AgCl NPs powder. The burn wound healing effect of CHI-AgCl NPs against CHI alone was examined, and Vaseline was used as a base component of the ointment. The composition of each ointment for burn wound healing is described in Table 1. The CHI and CHI-AgCl NPs powder were dissolved in distilled water to obtain 10% (w/v) aqueous solutions at room temperature, which were used in the water phase. Twelve grams of Vaseline, 12 g of stearyl alcohol and a 4 g of surfactant (Cremophor RH40, HCO-40, SIGMA, Darmstadt, Germany) were heated at 75 °C and mixed with the water phase solution (40 mL) at 75 °C to obtain an emulsion. Uniform ointments were obtained after slow cooling of the emulsion.

Table 1. The composition of each ointment for *in-vivo* burn wound healing in the rats.

Ointment Composition	Vaseline	CHI	CHI-AgCl NPs
Oil phase	12 g Vaseline, 12 g Stearyl alcohol, 4 g Cremophor RH40		
Water phase (40 mL)	–	4 g CHI	4 g CHI-AgCl NPs
Optical images			

2.4. Burn Wound Model

The experiment was approved by the institutional committee for animal care in laboratory research. For all experiments, four-week-old Sprague-Dawley rats were housed and bred at the experimental animal center of Chungnam National University. The animals were provided with a commercial diet and water *ad libitum* under temperature-, humidity-, and lighting-controlled conditions (22 ± 2 °C, $55 \pm 5\%$, and a 12:12-h light–dark cycle, respectively). Procedures involving animals and their care were conducted in accordance with our institutional guidelines, which comply with international laws and policies [25]. To induce burns with skin damage, a slightly modified soldering iron with a flat contact area of 28.3 mm^2 (AD = 6 mm) was made. Before creation of the burn wound, rats were anaesthetized by Tiletamine plus Zolazepam, (30 mg/kg + 10 mg/kg) according to body weight. Hair on the dorsal side of the rats was removed, and a burn wound was inflicted by placing the circular iron disc (heated to 95 °C) over the dorsal side for 20 s. Second-degree burns without cellular and tissue structure in the dermis were observed by sections stained with hematoxylin and eosin (H&E). Animals were divided after 7 days of acclimation in the cage, and then assigned equally (n = 30) to one of the following groups: Group 1-untreated controls dressed with gauze; Group 2-treated with the Vaseline ointment with gauze; Group 3-treated with CHI ointment with gauze; and group 4-treated with CHI-AgCl NPs ointment with gauze. Small amounts of ointment were applied everyday for 21 days. The sample tissue was enucleated after treatment 1, 3, 7, 14, and 21 days before sacrifice. The skin tissue for histopathological analysis was fixed in 10% buffered formalin, subsequently dehydrated, and embedded in paraffin. The tissue paraffin was cut into 5 μm sections. Fixed sections were then stained with H&E and Masson's trichrome staining (MT).

2.5. Analysis of Blood Counts

The biochemical analysis of the whole blood was performed to confirm the healing process. The blood sample was obtained from the abdominal aorta of animals before sacrifice. After being placed in a serum tube containing the EDTA, and the tube was shaken with an orbital shaker. Complete blood counts were measured using an automatic blood chemical analyzer (MS9-5, CARESIDE CO. LTD, Seongnam, Korea).

3. Results and Discussion

3.1. Characterization of CHI-AgCl NPs

The formation of CHI-AgCl NPs was confirmed easily by both the color change from yellow to yellowish brown with increasing reaction time and the strong surface plasmon resonance (SPR) peak around 400 nm due to the formation of spherical AgCl NPs [12]. Figure 1a shows some typical UV-Vis spectra of the suspension with CHI-AgCl NPs for the different reaction times. A strong SPR peak was observed at around 400 nm, which is a characteristic of spherical faceted AgCl NPs. A red shift of the absorption maximum was observed with increasing reaction time. The inset graph in Figure 1a shows the change in absorbance at $\lambda = 400$ nm of the CHI-AgCl NPs suspension with the reaction time for 300 min. To further investigate the morphology of CHI-AgCl NPs, the sample was analyzed with a field emission scanning electron microscope. There is a roughly spherical CHI-AgCl NPs, as seen in Figure 1b. The size of the roughly spherical CHI-AgCl NPs ranges between 30 and 50 nm, and their average diameter is 42 ± 15 nm (inset image in Figure 1b). It was assumed that the formation mechanism of AgCl NPs occurred in two steps. The mechanism on the formation of CHI-AgCl NPs is illustrated in Figure 1c. Firstly, the Ag ions readily reacted with Cl ions, which remained around the ammonium group of CHI after the acidic hydrolysis by hydrochloric acid (HCl). Secondly, small AgCl NPs were stabilized with amino and hydroxyl groups in CHI and underwent growth to large particles [26–30].

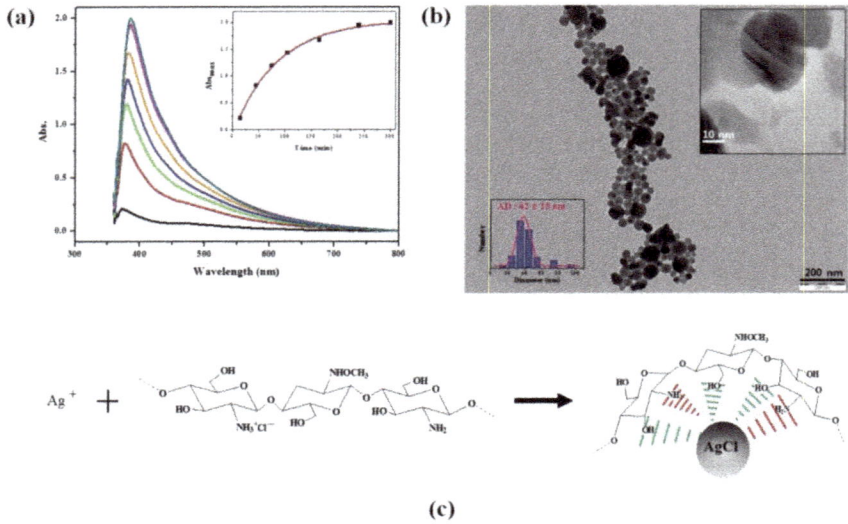

Figure 1. (**a**) UV-Vis spectra of CHI-AgCl NPs dispersion with reaction time (the inset shows the change in absorbance at 400 nm with reaction time); (**b**) TEM image and the corresponding particle size distribution (inset) of the CHI-AgCl NPs; and (**c**) a proposed mechanism for the formation of CHI-AgCl NPs.

3.2. A Clinical Pathology Study

A clinical pathology study was conducted during *in-vivo* burn wound treatment to determine the effects of burn trauma on skin that was infected and dehydrated after sustaining injury. The survival curves were found to be significantly different between the ointment-treated groups (Groups 2–4) and untreated group (Group 1). The survival rates of the CHI-AgCl NPs ointment group were 90%, but those of the untreated, Vaseline, and CHI groups were 80% at 21 days (n = 30). In all groups of animals, most of the fatalities occurred between Day 1 and 6 because of the post-burning infection (Figure 2a). In particular, the surviving rate was highest in the CHI-AgCl NPs ointment group in comparison with other ointment treated groups (Groups 2 and 3), making it clear that the CHI-AgCl NPs ointment had prevented the infection and improved the burn wound healing. The body weight gain and components of blood were determined for the clinical pathology study during the burn wound healing process. The body weight gain was compared in Figure 2b. On early treatment days until Day 3, the body weight gain was decreased for all groups. There was no significant difference in all groups during treatment days. White blood cells (WBC) are involved in defending the body against both infectious disease and foreign material. Although the normal range of a white blood cells was 6.60 to 12.6 m/mm^3, the amount of WBC was somewhat increased in all of the untreated and treated groups after sustaining injury [31]. It was still increased in the untreated group with increasing treatment days, but those of the other treated groups were decreased. The variation of WBC in CHI-AgCl NPs treated group (Group 4) was the lowest of all groups, which indicated that the CHI-AgCl NPs ointment reduced the infection, resulting in accelerating the burn wound healing. The amount of platelets was increased in the all groups during early-treatment days because the burn wound was created, and was reduced thereafter with increasing treatment days (Figure 2c,d).

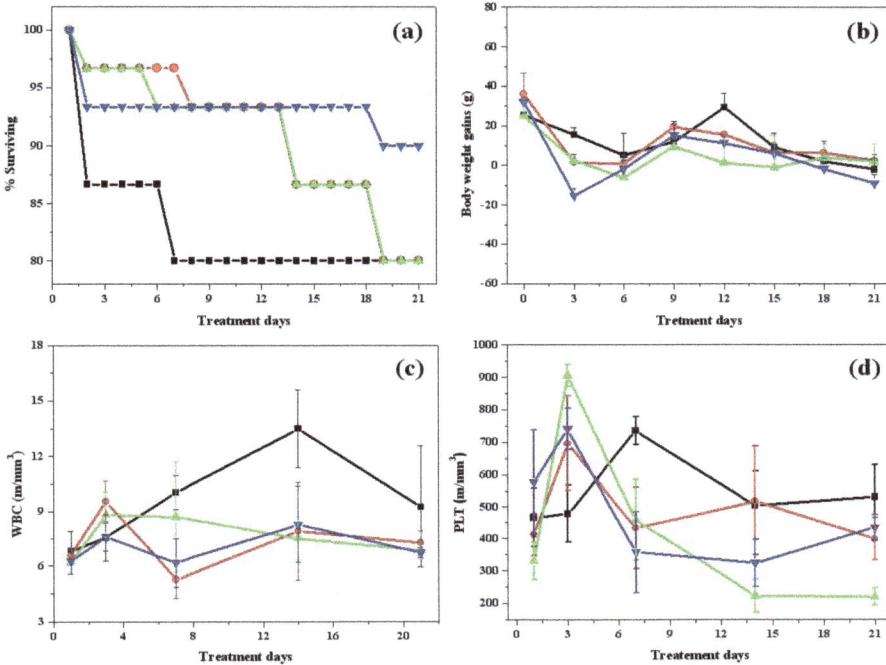

Figure 2. The *in-vivo* clinical pathology study of untreated (■), Vaseline (●), CHI (▲) and CHI-AgCl NPs ointment (▼) on: survival curves (**a**); body weight gains (**b**); concentration of white blood cells (WBC) (**c**); and concentration of platelet (PLT) (**d**).

3.3. Histological Analysis

The healing pattern of the burn wounds was studied to examine the histology for the untreated, Vaseline, CHI, and CHI-AgCl NPs groups on Days 1, 3, 7, 14, and 21 using H&E staining method (Figure 3). Generally, the injury initiates inflammatory phases for healing [32]. Therefore, it is difficult to assess whether the inflammatory response is part of the normal healing process or an effect of the material during the early stages of wound healing. For the untreated group, the epidermis is completely destroyed and interrupted, and the tissue of the wound is filled by necrotic material, bulla, and infiltrated inflammation until Day 21. Hyperkeratosis was also observed on Days 7 and 14. The Vaseline-treated group was shown to be similar to the untreated group. At 21 days, however, the Vaseline-treated group exhibited less infiltrated inflammation than untreated group, but not bulla. In the CHI ointment-treated group, the number of bulla was gradually decreased, and infiltrated inflammation was not observed in comparison to the Vaseline-treated group. In contrast, the CHI-AgCl NPs ointment-treated group was shown to have little infiltrated inflammation, and the underlying area showed fibrosis with proliferation of fibroblasts that were placed in the granulation tissue, and started to fill up the regenerated tissue from Day 14. After 21 days, the epidermis was filled with mature granulation tissue like normal tissue, confirming the accelerated burn wound healing by the improved antibacterial activity of the AgCl NPs stabilized with CHI.

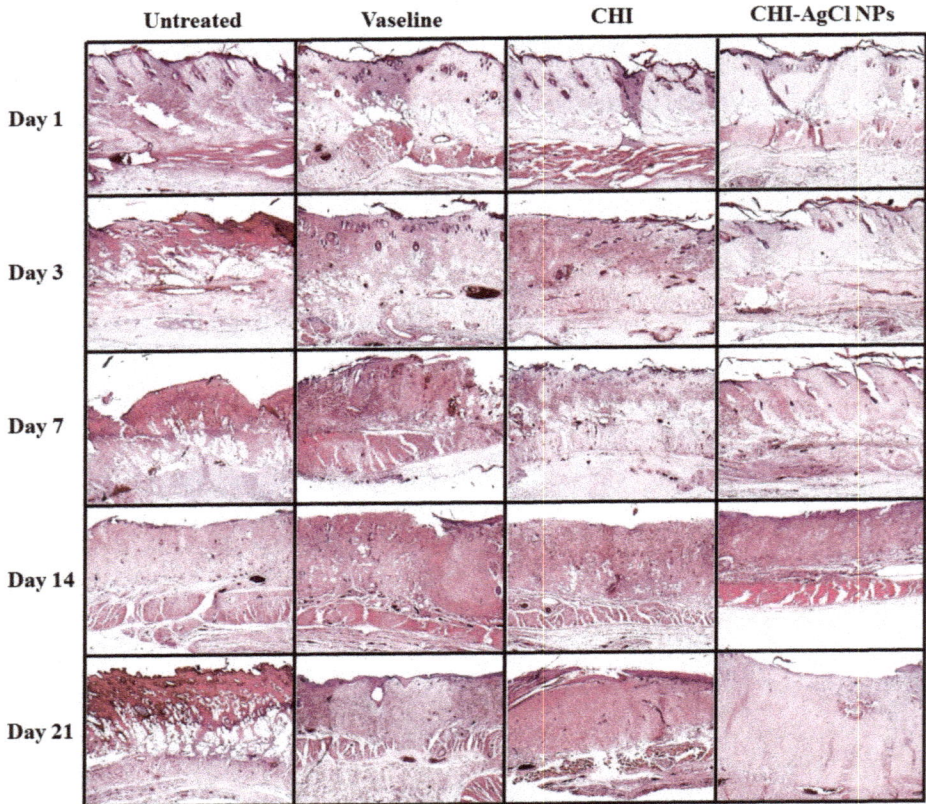

Figure 3. Evaluation of histopathology in the healing effect of ointments on burn induced skin damage using the H&E staining. The photographs were taken at an original magnification of ×40.

3.4. Evaluation of Collagen Percentage

Collagen is the main structural protein component of connective tissue, and is mainly found in skin. Therefore, it is desirable to observe the regenerated collagen in damaged skin during the healing process. To observe the collagenous components, the wounds with or without treated ointment were stained using a Masson's trichrome staining method. Figure 4 shows representative photomicrographs of wound healing on Days 3, 7 and 14. The collagenous components were stained in blue, and the cytoplasm appeared in varying shades of red. It was precisely observed that the necrosis was filled in the regenerated tissue in untreated group (Group 1), whereas the ointment-treatment (Groups 2–4) influenced the regeneration of collagen for burn wound healing. For an accurate description, the density of collagen at 14 days was determined using an image analyzer (Nikon, Japan), as shown in Figure 5. The relative collagen density was increased significantly in the ointment-treated groups, compared to that of the untreated group. Among them, the CHI-AgCl NPs-treated group (Group 4) showed the highest collagen density, suggesting increased collagenase activity by the dual antibacterial activity of the AgCl NPs stabilized with CHI. The data were represented as the mean ± standard deviation (SD) of 10 independent experiments, and difference was significant at $p < 0.05$ compared with the untreated, Vaseline, and CHI groups.

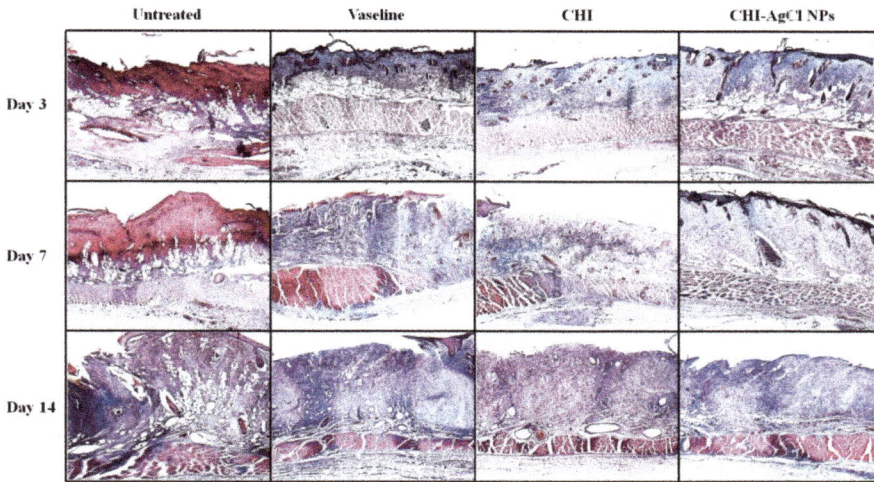

Figure 4. Evaluation of histopathology in the healing effect of ointment on burn induced sk.n damage using the Masson's trichrome staining. The photographs were taken at an original magnificat.on of ×40.

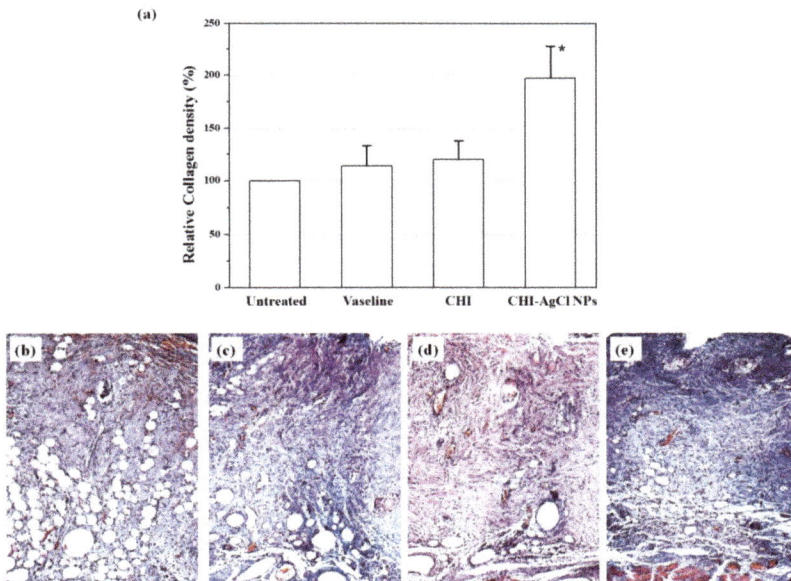

Figure 5. (**a**) The relative collagen density in burn induced skin damage at treatment Day 14. Evaluation of histopathology in the healing effect of untreated (**b**); Vaseline (**c**); CHI (**d**); and CHI-AgCl NPs (**e**) ointment using MT staining. The photographs were taken at an original magnification of ×100. The data are mean ± SD of 10 independent experiments. * indicates data with a statistical significance ($p < 0.05$) compared with the untreated and CHI-AgCl NPs ointment-treated groups.

4. Conclusions

In view of the rapid progress of application of nanomaterials in bioengineering fields, environmentally friendly methods should be required since the common methods generate toxic and biological hazards. The CHI-AgCl NPs was successfully obtained by a simple and environmentally benign method that used water and CHI as a biomaterial. The CHI is fundamental in the formation and stabilization of well-dispersed AgCl NPs with a mean diameter of 42 ± 15 nm. The burn wound treated with CHI-AgCl NPs ointment showed the highest percent of survival, and a reasonable number of white blood cells because of the prevention of infection in wound healing. The collagenous components were more regenerated in the CHI-AgCl NPs ointment group than the other treated groups. It was demonstrated that the CHI-AgCl NPs were more effective as a wound-healing accelerator because of improved antibacterial activity of CHI exerted by the AgCl NPs. Consequently, CHI-AgCl NPs ointment is considered a promising material for burn wound healing.

Acknowledgments: This research was supported by Basic Science Research Program through the National Research Foundation of Korea (NRF) funded by the Ministry of Science, ICT & Future Planning (2015M2A2A6A03044942).

Conflicts of Interest: The authors declare no conflict of interest.

References

1. Chaloupka, K.; Malam, Y.; Seifalian, A.M. Nanosilver as a new generation of nanoproduct in biomedical applications. *Trends Biotechnol.* **2010**, *28*, 580–588. [CrossRef] [PubMed]
2. Lee, S.J.; Heo, D.N.; Moon, J.H.; Ko, W.K.; Lee, J.B.; Bae, M.S.; Park, S.W.; Kim, J.E.; Lee, D.H.; Kim, E.C.; Lee, C.H.; Kwon, I.K. Electrospun chitosan nanofibers with controlled levels of silver nanoparticles. Preparation, characterization and antibacterial activity. *Carbohyd. Polym.* **2014**, *111*, 530–537. [CrossRef] [PubMed]
3. Atiyeh, B.S.; Costagliola, M.; Hayek, S.N.; Dibo, S.A. Effect of silver on burn wound infection control and healing: Review of the literature. *Burns* **2007**, *33*, 139–148. [CrossRef] [PubMed]
4. Cuttle, L.; Mill, J.; Kimble, R.M. Acticoat™: A cost-effective and evidence-based dressing strategy. *Burns* **2008**, *34*, 578–579. [CrossRef]
5. Silver, S.; Phung, L.T.; Silver, G. Silver as biocides in burn and wound dressings and bacterial resistance to silver compounds. *J. Ind. Microbiol. Biotechnol.* **2006**, *33*, 627–634. [CrossRef] [PubMed]
6. Huang, Y.; Li, X.; Liao, Z.; Zhang, G.; Liu, Q.; Tang, J.; Peng, Y.; Liu, X.; Luo, Q. A randomized comparative trial between Acticoat and SD-Ag in the treatment of residual burn wounds, including safety analysis. *Burns* **2007**, *33*, 161–166. [CrossRef] [PubMed]
7. Wright, J.B.; Lam, K.; Buret, A.G.; Olson, M.E.; Burrell, R.E. Early healing events in a porcine model of contaminated wounds: Effects of nanocrystalline silver on matrix metalloproteinases, cell apoptosis, and healing. *Wound Repair Regen.* **2002**, *10*, 141–151. [CrossRef] [PubMed]
8. Min, S.H.; Yang, J.H.; Kim, J.Y.; Kwon, Y.U. Development of white antibacterial pigment based on silver chloride nanoparticles and mesoporous silica and its polymer composite. *Micropor. Mesopor. Mat.* **2010**, *128*, 19–25. [CrossRef]
9. Li, X.; Zuo, W.; Luo, M.; Shi, Z.; Cui, Z.; Zhu, S. Silver chloride loaed hollow mesoporous aluminoilica spheres and their application in antibacterial coatings. *Mater. Lett.* **2013**, *105*, 159–161. [CrossRef]
10. Adams, A.; Santschi, E.; Mellencamp, M. Antibacterial properties of a silver chloride-coated nylon wound dressing. *Vet. Surg.* **1999**, *28*, 219–225. [CrossRef]
11. Gopinath, V.; Priyadarshini, S.; Priyadharsshini, N.M.; Pandian, K.; Velusamy, P. Biogenic synthesis of antibacterial silver chloride nanoparticles using leaf extracts of *Cissus quadrangularis* Linn. *Mater. Lett.* **2013**, *91*, 224–227. [CrossRef]
12. Youn, M.H.; Lim, Y.M.; Gwon, H.J.; Park, J.S.; An, S.J.; Nho, Y.C. Characterization of an antibacterial silver chloride/poly(acrylic acid) deodorant prepared by a gamma-ray irradiation. *Macromol. Res.* **2009**, *17*, 813–816. [CrossRef]
13. Barikani, M.; Oliaei, E.; Seddiqi, H.; Honarkar, H. Preparation and application of chitin and its derivatives: A review. *Iran. Polym. J.* **2014**, *23*, 307–326. [CrossRef]

14. Suzuki, K.; Mikami, T.; Okawa, Y.; Tokoro, A.; Suzuki, S.; Suzuki, M. Antitumor effect of hexa-N-acetylchitohexaose and chitohexaose. *Carbohyd. Res.* **1986**, *151*, 403–408. [CrossRef]

15. No, H.K.; Park, N.Y.; Lee, S.H.; Meyers, S.P. Antibacterial activity of chitosans and chitosan oligomers with different molecular weights. *Int. J. Food Microbiol.* **2002**, *74*, 65–72. [CrossRef]

16. Li, P.; Poon, Y.F.; Li, W.; Zhu, H.Y.; Yeap, S.H.; Cao, Y.; Qi, X.; Zhou, C.; Lamrani, M.; Beuerman, R.W. A polycationic antimicrobial and biocompatible hydrogel with microbe membrane suctioning ability. *Nat. Mater.* **2010**, *10*, 149–156. [CrossRef] [PubMed]

17. Rabea, E.I.; Badawy, M.E.T.; Stevens, C.V.; Smagghe, G.; Steurbaut, W. Chitosan as antimicrobial agent: Applications and mode of action. *Biomacromolecules* **2003**, *4*, 1457–1465. [CrossRef] [PubMed]

18. Busilacchi, A.; Gigante, A.; Mattioli-Belmonte, M.; Manzotti, S.; Muzzarelli, R.A. Chitosan stabilizes platelet growth factors and modulates stem cell differentiation toward tissue regeneration. *Carbohyd Polym.* **2013**, *98*, 665–676. [CrossRef] [PubMed]

19. Kojima, K.; Okamoto, Y.; Miyatake, K.; Fujise, H.; Shigemasa, Y.; Minami, S. Effects of chitin and chitosan on collagen synthesis in wound healing. *J. Vet. Med. Sci.* **2004**, *66*, 1595–1598. [CrossRef] [PubMed]

20. do Nascimento, E.G.; Sampaio, T.B.M.; Medeiros, A.C.; de Azevedo, E.P. Evaluation of chitosan gel with 1% silver sulfadiazine as an alternative for burn wound treatment in rats. *Acta Cir. Bras.* **2009**, *24*, 460–465. [CrossRef]

21. Santos, T.; Marques, A.; Silva, S.; Oliveira, J.M.; Mano, J.; Castro, A.G.; Reis, R. *In vitro* evaluation of the behaviour of human polymorphonuclear neutrophils in direct contact with chitosan-based membranes. *J. Biotechnol.* **2007**, *132*, 218–226. [CrossRef] [PubMed]

22. Ueno, H.; Murakami, M.; Okumura, M.; Kadosawa, T.; Uede, T.; Fujinaga, T. Chitosan accelerates the production of osteopontin from polymorphonuclear leukocytes. *Biomaterials* **2001**, *22*, 1667–1673. [CrossRef]

23. Kumar, M.R.; Muzzarelli, R.A.; Muzzarelli, C.; Sashiwa, H.; Domb, A. Chitosan chemistry and pharmaceutical perspectives. *Chem. Rev.* **2004**, *104*, 6017–6084. [CrossRef] [PubMed]

24. Kang, Y.O.; Lee, T.S.; Park, W.H. Green synthesis and antimicrobial activity of silver chloride nanoparticles stabilized with chitosan oligomer. *J. Mater. Sci. Mater. Med.* **2014**, *25*, 2629–2638. [CrossRef [PubMed]

25. Clark, J.D.; Gebbart, G.F.; Gonder, J.C.; Keeling, M.E.; Kohn, D.F. *Guide for the Care and Use of Laboratory Animals*, 6th ed.; National Institutes of health (NIH): Bethesda, MD, USA, 1985.

26. Croisier, F.; Jérôme, C. Chitosan-based biomaterials for tissue engineering. *Eur. Polym. J.* **2013**, *49*, 780–792. [CrossRef]

27. Kadib, A.E.; Molvinger, K.; Cacciaguerra, T.; Bousmina, M.; Brunel, D. Chitosan templated synthesis of porous metal oxide microspheres with filamentary nanostructures. *Micropor. Mesopor. Mater.* **2011**, *142*, 301–307. [CrossRef]

28. Kramareva, N.V.; Stakheev, A.Y.; Tkachenko, O.P.; Klementiev, K.V.; Grünert, W.; Finashina, E.D.; Kustov, L.M. Heterogenized palladium chitosan complexes as potential catalysts in oxidation reactions: Study of the structure. *J. Mol. Catal. A Chem.* **2004**, *209*, 97–106. [CrossRef]

29. Regiel, A.; Irusta, S.; Kyzioł, A.; Arruebo, M.; Santamaria, J. Preparation and characterization of chitosan–silver nanocomposite films and their antibacterial activity against *Staphylococcus aureus*. *Nanotechnology* **2013**, *24*, 015101:1–015101:13. [CrossRef] [PubMed]

30. Travan, A.; Pelillo, C.; Donati, I.; Marsich, E.; Benincasa, M.; Scarpa, T.; Semeraro, S.; Turco G.; Gennaro, R.; Paoletti, S. Non-cytotoxic silver nanoparticle-polysaccharide nanocomposites with antimicrobial activity. *Biomacromolecules* **2009**, *10*, 1429–1435. [CrossRef] [PubMed]

31. Hong, E.G.; Kim, M.G. Effect of Polysaccharide from Schizophyllum commune on burn and wound healing. *Korean Chem. Eng. Res.* **2006**, *44*, 87–91.

32. Balakrishnan, B.; Mohanty, M.; Umashankar, P.; Jayakrishnan, A. Evaluation of an in situ forming hydrogel wound dressing based on oxidized alginate and gelatin. *Biomaterials* **2005**, *26*, 6335–6342. [CrossRef] [PubMed]

materials

MDPI

Article

Effect of Graphene Nanoplatelets on the Physical and Antimicrobial Properties of Biopolymer-Based Nanocomposites

Roberto Scaffaro *, Luigi Botta, Andrea Maio, Maria Chiara Mistretta
and Francesco Paolo La Mantia

Dipartimento di Ingegneria Civile, Ambientale, Aerospaziale, dei Materiali, Università di Palermo,
UdR INSTM di Palermo, Viale delle Scienze, Palermo 90128, Italy; luigi.botta@unipa.it (L.B.);
andrea.maio@unipa.it (A.M.); mariachiara.mistretta@unipa.it (M.C.M.);
francescopaolo.lamantia@unipa.it (F.P.L.M.)
* Correspondence: roberto.scaffaro@unipa.it; Tel.: +39-091-23863723

Academic Editor: Mauro Pollini
Received: 5 April 2016; Accepted: 2 May 2016; Published: 9 May 2016

Abstract: In this work, biopolymer-based nanocomposites with antimicrobial properties were prepared via melt-compounding. In particular, graphene nanoplatelets (GnPs) as fillers and an antibiotic, *i.e.*, ciprofloxacin (CFX), as biocide were incorporated in a commercial biodegradable polymer blend of poly(lactic acid) (PLA) and a copolyester (BioFlex®). The prepared materials were characterized by scanning electron microscopy (SEM), and rheological and mechanical measurements. Moreover, the effect of GnPs on the antimicrobial properties and release kinetics of CFX was evaluated. The results indicated that the incorporation of GnPs increased the stiffness of the biopolymeric matrix and allowed for the tuning of the release of CFX without hindering the antimicrobial activity of the obtained materials.

Keywords: nanocomposites; graphene nanoplatelets (GnPs); poly(lactic acid) (PLA); antimicrobial activity; drug release; ciprofloxacin

1. Introduction

Biopolymers are an alternative to oil-based synthetic polymers since they are renewable and do not contribute to environmental pollution being biodegradable, and they are currently used in several applications [1–6]. Nevertheless, a broad use of biopolymers is often restricted by the necessity of improving some functional properties such as mechanical and barrier properties. Therefore, intense efforts have been made to improve their physical properties in order to enhance the commercial potential of biopolymers such as poly(lactic acid) (PLA) [7–11].

An effective way to improve the properties of biopolymers is to incorporate nano-sized reinforcements in the matrix [12–16]. Indeed, it is well known that nanocomposites, *i.e.*, polymers filled with particles having at least one dimension in the nanoscale range, show unique properties because of the peculiar increase of the matrix-filler interface [17]. Various nano-sized fillers, such as layered silicates, metal, polyhedral oligomeric silsesquioxane, carbon nanomaterials, and silica nanoparticles, are being developed and extensively studied in the field of polymer nanocomposites. Among the above materials, graphene, a flat monolayer of sp^{2-} bonded carbon atoms, is very promising due to its unique characteristics such as high electronic conductivity, large specific surface area, and high mechanical strength. Graphene combines a layered structure of clays with superior mechanical and thermal properties of carbon nanotubes, which can provide excellent functional property enhancements.

Recently, many studies have been conducted to provide biopolymers with antimicrobial properties that might encourage their use in the field of active food packaging as well as in specific biomedical applications [18–23].

Providing a polymeric device with antibacterial properties can be reached by different routes, with or without the modification of the polymer structure. The incorporation of antimicrobials, or other molecules, into a polymer matrix via melt-processing is a method that has been widely adopted in the recent past since it has the advantage of using equipment commonly and already used to process thermoplastic materials [2,3,24–31]. This method, moreover, ensures large production volumes and solventless systems with obvious positive economic and environmental implications.

The effectiveness of the antimicrobial activity over time is mainly determined by the release rate of the antimicrobial compounds. Release kinetics that are either too slow or too fast must be avoided since the former means that microbial growth is not sufficiently inhibited and the latter means that inhibition will not be sustained over time. The rate of release depends on different factors, *i.e.*, the type of polymeric matrix, the preparation method, the environmental conditions, the interactions between the antimicrobial, and the matrix. In this regard, nanoparticles can potentially be used to control the release of antimicrobial agents as reported in the scientific literature [17,32].

The aim of this work was to prepare and characterize biopolymer-based nanocomposites with antimicrobial properties suitable for medical device packaging. In particular, graphene nanoplatelets (GnPs) as fillers and ciprofloxacin (CFX) as biocide were incorporated via melt-compounding in a commercial biodegradable polymer blend (BioFlex®) based on PLA. CFX is a wide-spectrum antibiotic belonging to the fluoroquinolone family, active against both Gram-negative and Gram-positive strains. Its spectrum of activity includes most strains of bacterial pathogens responsible for respiratory, urinary tract, gastrointestinal, and abdominal infections.

The rheological, mechanical, and antimicrobial properties of the obtained nanocomposites were evaluated. In particular, the influence of GnPs on the antimicrobial properties and release kinetics was studied.

2. Materials and Methods

2.1. Materials

The polymer matrix used in this work was a sample of a biodegradable polymer blend of proprietary composition (Bioflex), trade name Bio-Flex® F2110, supplied by FKUR (Willich, Germany). It is based on PLA and a biodegradable copolyester and contains some additives.

Graphene nanoplatelets (GnPs), trade name xGnP®, Grade C, were supplied by XG Sciences Inc., Lansing, MI, USA. Each particle consists of several sheet of graphene with an average thickness of approximately 10–20 nm, average diameter between 1 and 2 μm, and a specific surface area of about 750 m^2/g.

Ciprofloxacin (CFX, chemical formula: $C_{17}H_{18}FN_3O_3$, T_m = 253–257 °C) was supplied by Sigma Aldrich (St. Louis, MO, USA) and used as received without further purification. Its water solubility is 80 mg/L at 30 °C and 120 mg/L at 40 °C [33].

2.2. Preparation of Nanocomposites and the Incorporation of CFX

GnPs were added to Bioflex at 1 wt % and 5 wt % at the solid state, and the mixtures have then melt-compounded in a batch mixer (Brabender PLE330, Duisburg, Germany) at 170 °C and a rotational speed of 60 rpm for 5 min. For comparison, a pristine polymer-blend matrix was processed under the same conditions. Before processing, Bioflex and GnPs were dried under vacuum at 90 °C for 4 h and at 120 °C overnight, respectively.

The incorporation of CFX was achieved via melt-compounding using the same batch mixer described above at the same processing conditions, *i.e.*, 170 °C and a rotational speed of 60 rpm. In detail, both the polymer and the GnPs were first fed to the mixer and compounded for 4 min.

Thereafter, the CFX was added and the blend was processed for no longer than 1 min in order to avoid eventual degradation phenomena of the additive. Both CFX and GnPs were added to the polymeric matrix at 5% (w/w). For comparison, Bioflex, incorporating only CFX at 5% (w/w), was processed under the same conditions. In Table 1, the composition of all the samples and their identification codes are reported.

Films were prepared by compression-molding using a laboratory press (Carver, Wabash, IN, USA). The material was preventively ground, placed in a mold between two Teflon sheets, and pressed at 170 °C and 100 bar for about 2 min to obtain a 200-μm-thick film.

Table 1. Composition of samples and their codes.

Sample Code	Bio-Flex® F2110 (BIO) (w/w %)	Graphene Nanoplatelets (GnPs) (w/w %)	Ciprofloxacin (CFX) (w/w %)
BIO	100	–	–
BIO/GnP-1	99	1	–
BIO/GnP-5	95	5	–
BIO/CFX	95	–	5
BIO/GnP-5/CFX	90	5	5

2.3. Characterizations

The morphology of all the materials, including neat GnPs and neat CFX, was analyzed by scanning electron microscopy (SEM; Quanta 200 ESEM, FEI, Hillsboro, OR, USA). In particular, both the GnPs powder and the CFX powder were directly glued onto a sample holder, whereas the polymeric samples were fractured under liquid nitrogen and then glued onto a sample holder. All the samples were sputter-coated with a thin layer of gold to avoid electrostatic charging under the electron beam.

The rheological characterization was performed using a plate–plate rotational rheometer (HAAKE MARS, Thermo Scientific, Waltham, MA, USA), operating at 170 °C on samples obtained by compression-molding. The instrument has been set to operate in the frequency sweep mode in the range 0.1–500 rad/s with a strain of 5%. Before testing, the samples were dried for 4 h under vacuum at 90 °C.

Tensile mechanical measurements were carried out using a dynamometer (Instron model 3365, High Wycombe, UK) on rectangular shaped specimens (10 × 90 mm) cut off from films prepared by compression-molding as described above. The grip distance was 30 mm, and the crosshead speed was 5 mm/min.

The antimicrobial activity of the materials was determined by the agar diffusion method evaluating the presence of inhibition zones. In particular, *Micrococcus luteus* ATCC 10240 (ATCC, Manassas, VA, USA) was used as a tester strain in order to study antimicrobial property of the prepared materials. A bacterial suspension of ~10^9 colony forming units (CFU) was inoculated into 5 mL of lysogeny broth (LB)-soft agar to obtain a uniform bacterial overlay on LB-agar plates. Circular samples (diameter 12 mm) containing CFX at 5% (w/w) were placed over a bacterial tester overlay. Samples without CFX, *i.e.*, BIO and BIO/GnP-5, were used as controls. Bacterial growth inhibition halos were observed after overnight incubation at 37 °C.

A series of CFX solutions of distilled water containing 0.1–5 mg/L of CFX was used to obtain a calibration curve correlating the absorbance peak intensity and the CFX concentration using a UV/vis spectrophotometer (model UVPC 2401, Shimadzu Italia s.r.l., Milan, Italy). In the concentration range here investigated, the calibration curve was found to be a line. The maximum absorbance peak of CFX was detected at 276 nm. The release of CFX from the films was investigated by immersing a pre-weighed sample in 10 mL of distilled water stored at 37 °C. At specific time intervals, for 6 weeks, the absorbance peak intensity at 276 nm of the storage solutions was measured and converted to the quantities of CFX released based on the calibration line previously calculated. After each measurement, the samples were immersed in 10 mL of fresh distilled water and the cumulative release of CFX here reported was calculated by sequentially adding the CFX released after each step. Each measurement was performed in triplicate.

3. Results and Discussion

3.1. Properties of Nanocomposites

SEM micrographs of the neat polymeric matrix, of the GnPs, and of the nanocomposite systems are reported in Figure 1. The surface fracture of neat BIO (Figure 1a) exhibited the typical morphology of polymer blends showing a poor interfacial adhesion, evidenced by the presence of voids due to the detachment of the copolyester phase from the PLA matrix during the sample breaking. The SEM micrograph of neat GnPs powder (Figure 1b) revealed irregular aggregates having different dimensions. BIO/GnP-1 (Figure 1c) showed a fairly good dispersion of the GnPs and a good adhesion between the matrix and the filler. As expected, upon increasing the filler concentration, the aggregates visible in the fracture section increased, although a good adhesion was still evident (Figure 1d).

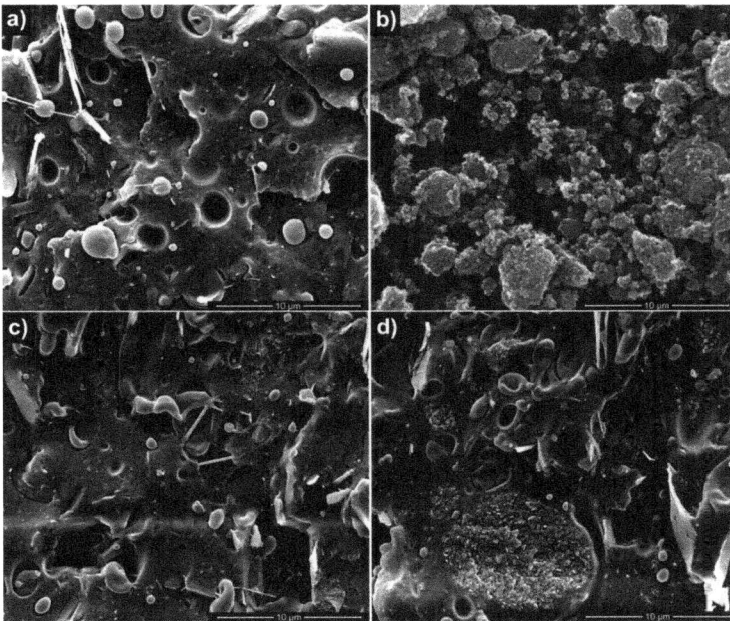

Figure 1. SEM micrographs of: (**a**) BIO; (**b**) neat GnP; (**c**) BIO/GnP-1; (**d**) BIO/GnP-5.

The influence of GnPs on the rheological behavior of the obtained systems was evaluated. The complex viscosity values as a function of frequency are reported in Figure 2 for neat BIO and for the related nanocomposites. The viscosity of the samples filled with GnPs was higher than that of the neat matrix, and it increased with increasing the filler content. Moreover, both BIO/GnP-1 and BIO/GnP-5 exhibited a more pronounced non-Newtonian behavior if compared with the unfilled polymer. At high frequencies, the flow curves of the nanocomposite samples and of the unfilled matrix were much closer among them, although the viscosity values of the nanocomposites remained higher than that of the neat BIO. This rheological behavior is reported in the scientific literature as a typical behavior shown by several nanocomposite systems including polymer/clay nanocomposites [34,35] and polymer/GnPs nanocomposites [36,37]. It is generally attributed to a strong interaction between the dispersed filler and the matrix that restricts the polymer chain movements.

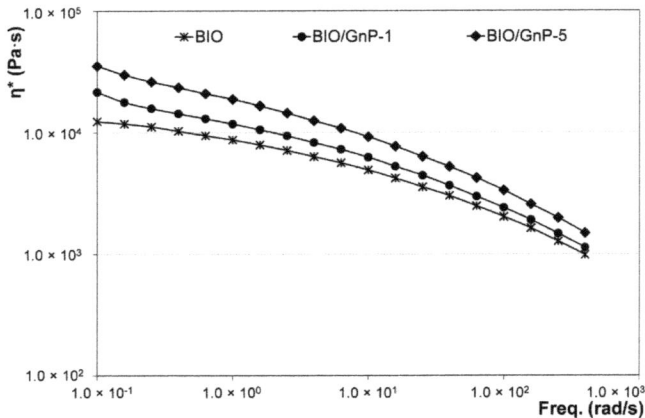

Figure 2. Complex viscosity as a function of frequency of neat BIO and related nanocomposites.

In Table 2, the elastic modulus (E), the tensile strength (TS), and the elongation at break (EB) of neat BIO and of the nanocomposite systems are reported. The adding of GnPs provoked an increase of the rigidity of the biodegradable matrix. In particular, the modulus increased on increasing the amount of filler and BIO/GnP-5 showed a tensile modulus about 40% higher than that of neat BIO. Tensile strength and elongation at break decreased in the presence of the nanoparticles. In detail, the decrease of these properties was very slight for BIO/GnP-1, whereas adding the 5% of GnPs led to a higher decrement, especially of the EB. The decrease of the tensile strength can be considered a consequence of the reduction of the elongation at break that was likely due to the presence of some GnPs aggregates, as shown by SEM micrographs. However, the reduction is quite low and does not compromise the use of the material.

Table 2. Elastic modulus (E), tensile strength (TS), and elongation at break (EB) of neat BIO and related nanocomposites.

Samples	E (MPa)	TS (MPa)	EB (%)
BIO	140 ± 5.8	11.6 ± 0.5	165 ± 7.2
BIO/GnP-1	168 ± 6.2	11.1 ± 0.4	151 ± 6.1
BIO/GnP-5	195 ± 6.8	10.6 ± 0.6	135 ± 6.9

3.2. Properties of Antimicrobial Nanocomposites

Following the results described above regarding the effect of GnP amount on properties of nanocomposites, CFX was incorporated in the nanocomposite system containing the 5% of GnPs. For comparison, BIO, incorporating only the CFX, was prepared. In Figure 3, the SEM micrographs of neat CFX, BIO/CFX, and BIO/GnP-5/CFX are reported. The micrograph of neat CFX powder (Figure 3a) showed that CFX is visible under the form of irregular crystalline aggregates formed by needle-like crystals. These aggregates were clearly visible in the cross section of the films containing CFX as shown by SEM micrographs reported in Figure 3b,c. In particular, the CFX was well dispersed both in the neat polymeric matrix (Figure 3b) and in the nanocomposite system (Figure 3c) in which both the CFX and the GnPs were clearly visible. Moreover, in both cases, it is evident that the dimensions of the CFX aggregates are smaller than the initial ones. This change in size can be attributed to the melt-compounding that caused a disaggregation of the CFX clusters and the dispersion of the particles into the matrix. Nevertheless, the adhesion level between CFX crystals and the polymer matrix is quite poor.

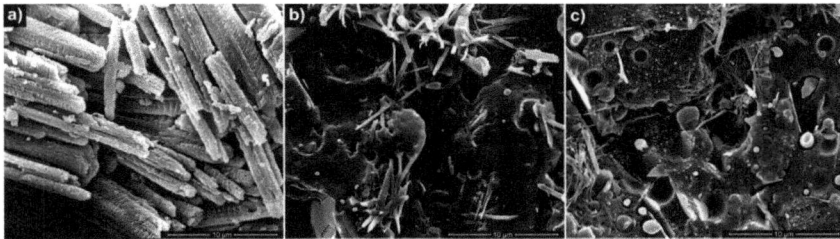

Figure 3. SEM micrographs of: (**a**) neat CFX; (**b**) BIO/CFX; (**c**) BIO/GnP-5/CFX.

The rheological behavior of the systems incorporating the antibiotic was evaluated and compared with the respective systems without CFX. In Figure 4, the complex viscosity values as a function of frequency of BIO/CFX and BIO/GnP-5/CFX were reported and compared with the flow curves of BIO and BIO/GnP-5 shown above.

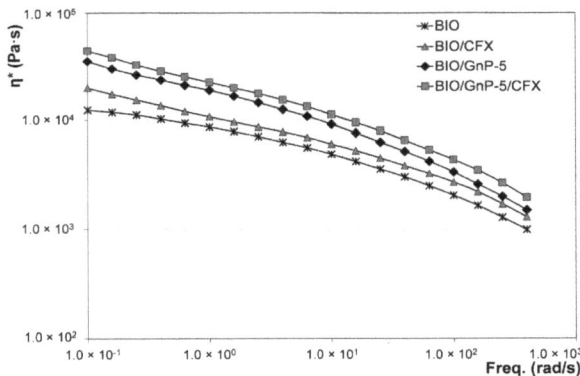

Figure 4. Complex viscosity as a function of frequency of systems incorporating CFX and of the respective systems without antimicrobial additive.

The presence of CFX led to a slight rising of the flow curves both of the neat biopolymeric matrix and of the nanocomposite system. Indeed, CFX acted as a micro-filler, causing the light increment of the viscosity in the whole range of investigated frequencies.

In order to verify whether the CFX incorporation caused some modification of the mechanical performance of the materials, tensile tests were performed. Table 3 reports the elastic modulus (E), the tensile strength (TS), and the elongation at break (EB) of all the materials containing CFX. The results showed that, for both systems, *i.e.*, BIO/CFX and BIO/GnP-5/CFX, adding the antimicrobial additive led to a slight increase of E and a decrease of EB. Indeed, CFX acted as a micro-filler, causing, at this concentration, a slight increment of the rigidity and decrement of the ductility.

Table 3. Elastic modulus (E), tensile strength (TS), and elongation at break (EB) of BIO/CFX and BIO/GnP-5/CFX.

Samples	E (MPa)	TS (MPa)	EB (%)
BIO/CFX	150 ± 5.1	10.4 ± 0.5	130 ± 9.7
BIO/GnP-5/CFX	206 ± 7.1	10.2 ± 0.6	121 ± 9.9

To verify that the incorporation of CFX conferred antimicrobial activity to the polymeric systems, agar diffusion tests were performed. A Gram-positive bacterium, *i.e.*, *M. luteus*, was used as tester strain to evaluate the growth inhibition zone around the samples. Figure 5 reports the bacterial inhibition halos observed around the systems incorporating the CFX and, for comparison, the respective systems without antimicrobial.

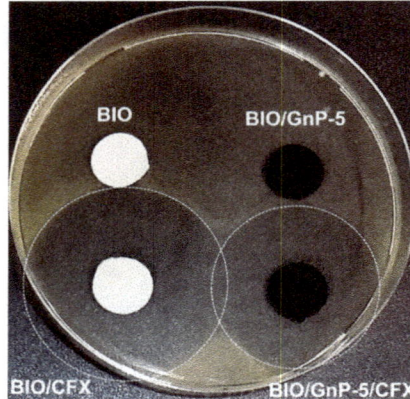

Figure 5. Agar diffusion test performed on *M. luteus* overlay for the materials incorporating CFX and for the respective systems without antimicrobial additive.

As expected, both the neat BIO and BIO-GnP-5 showed no antibacterial activity. On the contrary, large bacterial growth inhibition halos were observed around both samples incorporating CFX after an overnight incubation at 37 °C. More specifically, the presence of GnPs led to a reduction of the inhibition zone, *i.e.*, the inhibition halo diameter of BIO/CFX was about 42 mm, whereas BIO/GnP-5/CFX exhibited an inhibition halo diameter of about 35 mm. This result can be explained considering that the antimicrobial properties of the films are dependent on the release of CFX from themselves and that the presence of the GnPs likely influenced the CFX release from the sample.

In order to verify this hypothesis, the release kinetics in distilled water at 37 °C was evaluated. In Figure 6, cumulative CFX release as a function of the time from BIO/CFX and BIO/GnP-5/CFX are reported. In particular, the release data were expressed as M_t/M_∞, where M_t is the cumulative amount of drug released at time t, and M_∞ is the cumulative amount of drug released at infinite time (which should be equal to the theoretical absolute amount of drug incorporated within the system at time t = 0).

For both systems, the release of CFX was characterized by an initial burst phase followed by a second phase, which is characterized by a slower release rate. However, according to results of the agar diffusion test, BIO/GnP-5 released a lower amount of CFX during the entire six weeks of immersion. This result can be explained considering that the incorporated GnPs into the polymer matrix was able to create a tortuous pathway, thus slowing down the diffusion of drug molecules through the polymeric matrix, as already reported in the scientific literature for other nanocomposite systems [17,32].

In order to understand the release mechanism, the experimental data were fitted using the well-known power law model:

$$\frac{M_t}{M_\infty} = kt^n \tag{1}$$

where k is a kinetic constant related to the properties of the drug delivery system, and n is the diffusion exponent that characterizes the release mechanism. In particular, when the value of n is $\leqslant 0.5$, it indicates that the drug release follows the Fickian diffusion mechanism [38,39]. The power trend-lines fitting the experimental data (red dashed lines) are reported in Figure 6 together with the related equations and the R^2 values. It is worth noting that the power law model well fitted the experimental release data since R^2 values were very high, *i.e.*, 0.9977 for both systems. The n values obtained by the fitting were 0.3171 and 0.3547 for BIO/CFX and BIO/GnP-5/CFX, respectively. This implies that the release of CFX from both systems followed a diffusion-controlled mechanism. The k value was generally related to the release kinetics, *i.e.*, a higher k value indicates a faster release. As expected, the value of k was higher for the release from BIO/CFX in comparison with the k value obtained from the release profile of BIO/GnP-5/CFX.

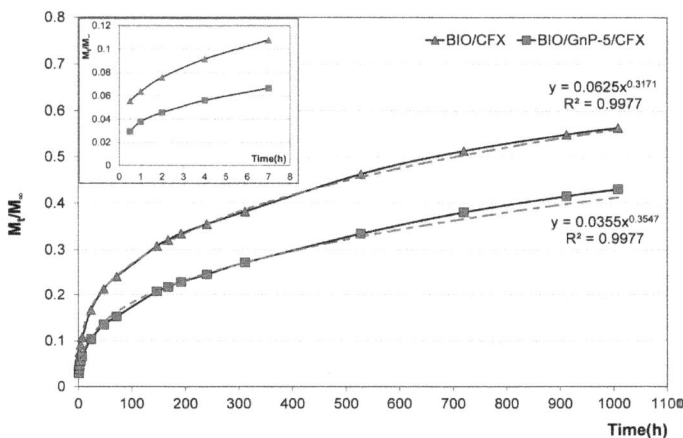

Figure 6. Cumulative CFX release from BIO/CFX and BIO/GnP-5/CFX. In the inset, only the first hours of release are plotted in order to enhance their readability. The power trend-lines fitting the experimental data (red dashed lines) are reported in the graph together with the respective equations and the R^2 values.

4. Conclusions

Biopolymer-based nanocomposites with antimicrobial properties filled with graphene nanoplatelets (GnPs) were prepared via melt-compounding in a batch mixer. An antibiotic, *i.e.*, ciprofloxacin (CFX), was chosen as biocide and incorporated together with GnPs in a commercial biodegradable polymer-blend matrix.

The morphological analysis revealed that the GnPs were well dispersed in the biodegradable matrix, although at the higher concentration some aggregates were visible. The nanocomposites exhibited flow curves higher than that of the neat Bioflex, and the viscosity increased as the filler content was increased. The adding of GnPs improved the stiffness of the matrix—in particular, the elastic modulus increased with an increasing filler amount. The incorporation of GnPs affected the release of CFX without hindering the antimicrobial activity of the obtained materials. In particular, the presence of GnPs led to a slower release of CFX.

Acknowledgments: The authors thank Anna Maria Puglia and Giuseppe Gallo (Dipartimento di Scienze e Tecnologie Biologiche Chimiche e Farmaceutiche, Università di Palermo) for performing agar diffusion tests. This work has been financially supported by Consorzio Interuniversitario Nazionale per la Scienza e Tecnologia dei Materiali (INSTM).

Author Contributions: Roberto Scaffaro and Luigi Botta conceived and designed the experiments; Luigi Botta prepared all the materials and performed rheological and release measurements; Andrea Maio performed the mechanical characterization; Maria Chiara Mistretta performed the SEM analysis; all the authors analyzed the data; Roberto Scaffaro and Luigi Botta wrote the paper.

Conflicts of Interest: The authors declare no conflict of interest.

References

1. Van de Velde, K.; Kiekens, P. Biopolymers: Overview of several properties and consequences on their applications. *Polym. Test.* **2002**, *21*, 433–442. [CrossRef]
2. Scaffaro, R.; Botta, L.; Sanfilippo, M.; Gallo, G.; Palazzolo, G.; Puglia, A.M. Combining in the melt physical and biological properties of poly(caprolactone) and chlorhexidine to obtain antimicrobial surgical monofilaments. *Appl. Microbiol. Biotechnol.* **2013**, *97*, 99–109. [CrossRef] [PubMed]
3. Scaffaro, R.; Botta, L.; Gallo, G.; Puglia, A.M. Influence of Drawing on the Antimicrobial and Physical Properties of Chlorhexidine-Compounded Poly(caprolactone) Monofilaments. *Macromol. Mater. Eng.* **2015**, *300*, 1268–1277. [CrossRef]
4. Scaffaro, R.; Lopresti, F.; Botta, L.; Rigogliuso, S.; Ghersi, G. Preparation of three-layered porous PLA/PEG scaffold: Relationship between morphology, mechanical behavior and cell permeability. *J. Mech. Behav. Biomed. Mater.* **2016**, *54*, 8–20. [CrossRef] [PubMed]
5. La Mantia, F.P.; Botta, L.; Morreale, M.; Scaffaro, R. Effect of small amounts of poly(lactic acid) on the recycling of poly(ethylene terephthalate) bottles. *Polym. Degrad. Stab.* **2012**, *97*, 21–24. [CrossRef]
6. Scarfato, P.; Di Maio, L.; Incarnato, L. Recent advances and migration issues in biodegradable polymers from renewable sources for food packaging. *J. Appl. Polym. Sci.* **2015**, *48*. [CrossRef]
7. Fiore, V.; Botta, L.; Scaffaro, R.; Valenza, A.; Pirrotta, A. PLA based biocomposites reinforced with Arundo donax fillers. *Compos. Sci. Technol.* **2014**, *105*, 110–117. [CrossRef]
8. Botta, L.; Fiore, V.; Scalici, T.; Valenza, A.; Scaffaro, R. New Polylactic Acid Composites Reinforced with Artichoke Fibers. *Materials* **2015**, *8*, 7770–7779. [CrossRef]
9. Botta, L.; Mistretta, M.C.; Palermo, S.; Fragalà, M.; Pappalardo, F. Characterization and Processability of Blends of Polylactide Acid with a New Biodegradable Medium-Chain-Length Polyhydroxyalkanoate. *J. Polym. Environ.* **2015**, *23*, 478–486. [CrossRef]
10. Scaffaro, R.; Morreale, M.; Mirabella, F.; La Mantia, F.P. Preparation and Recycling of Plasticized PLA. *Macromol. Mater. Eng.* **2011**, *296*, 141–150. [CrossRef]
11. Scarfato, P.; Avallone, E.; Galdi, M.R.; Maio, L.; Incarnato, L. Preparation, characterization, and oxygen scavenging capacity of biodegradable α-tocopherol/PLA microparticles for active food packaging applications. *Polym. Compos.* **2015**. [CrossRef]
12. Bordes, P.; Pollet, E.; Avérous, L. Nano-biocomposites: Biodegradable polyester/nanoclay systems. *Prog. Polym. Sci.* **2009**, *34*, 125–155. [CrossRef]
13. Raquez, J.-M.; Habibi, Y.; Murariu, M.; Dubois, P. Polylactide (PLA)-based nanocomposites. *Prog. Polym. Sci.* **2013**, *38*, 1504–1542. [CrossRef]
14. Scaffaro, R.; Botta, L.; Passaglia, E.; Oberhauser, W.; Frediani, M.; Di Landro, L. Comparison of different processing methods to prepare poly(lactid acid)-hydrotalcite composites. *Polym. Eng. Sci.* **2014**, *54*, 1804–1810. [CrossRef]
15. La Mantia, F.P.; Mistretta, M.C.; Scaffaro, R.; Botta, L.; Ceraulo, M. Processing and characterization of highly oriented fibres of biodegradable nanocomposites. *Compos. Part B Eng.* **2015**, *78*, 1–7. [CrossRef]
16. Di Maio, L.; Scarfato, P.; Milana, M.R.; Feliciani, R.; Denaro, M.; Padula, G.; Incarnato, L. Bionanocomposite polylactic acid/organoclay films: Functional properties and measurement of total and lactic acid specific migration. *Packag. Technol. Sci.* **2014**, *27*, 535–547. [CrossRef]
17. Paul, D.R.; Robeson, L.M. Polymer nanotechnology: Nanocomposites. *Polymer* **2008**, *49*, 3187–3204. [CrossRef]
18. Martínez-Sanz, M.; Bilbao-Sainz, C.; Du, W.X.; Chiou, B.S.; Williams, T.G.; Wood, D.F.; Imam, S.H.; Orts, W.J.; Lopez-Rubio, A.; Lagaron, J.M. Antimicrobial Poly(lactic acid)-Based Nanofibres Developed by Solution Blow Spinning. *J. Nanosci. Nanotechnol.* **2015**, *15*, 616–627. [CrossRef] [PubMed]
19. Tawakkal, I.S.M.A.; Cran, M.J.; Miltz, J.; Bigger, S.W. A review of poly(lactic acid)-based materials for antimicrobial packaging. *J. Food Sci.* **2014**, *79*, R1477–R1490. [CrossRef] [PubMed]
20. Xu, X.; Yang, Q.; Wang, Y.; Yu, H.; Chen, X.; Jing, X. Biodegradable electrospun poly(L-lactide) fibers containing antibacterial silver nanoparticles. *Eur. Polym. J.* **2006**, *42*, 2081–2087. [CrossRef]
21. Jin, T.; Zhang, H. Biodegradable polylactic acid polymer with nisin for use in antimicrobial food packaging. *J. Food Sci.* **2008**, *73*, M127–M134. [CrossRef] [PubMed]

22. Liu, L.S.; Finkenstadt, V.L.; Liu, C.K.; Jin, T.; Fishman, M.L.; Hicks, K.B. Preparation of poly(lactic acid) and pectin composite films intended for applications in antimicrobial packaging. *J. Appl. Polym. Sci.* **2007**, *106*, 801–810. [CrossRef]

23. Toncheva, A.; Paneva, D.; Manolova, N.; Rashkov, I. Electrospun poly(L-lactide) membranes containing a single drug or multiple drug system for antimicrobial wound dressings. *Macromol. Res.* **2011**, *19*, 1310–1319. [CrossRef]

24. Douglas, P.; Andrews, G.; Jones, D.; Walker, G. Analysis of *in vitro* drug dissolution from PCL melt extrusion. *Chem. Eng. J.* **2010**, *164*, 359–370. [CrossRef]

25. Nostro, A.; Scaffaro, R.; Ginestra, G.; D'Arrigo, M.; Botta, L.; Marino, A.; Bisignano, G. Control of biofilm formation by poly-ethylene-co-vinyl acetate films incorporating nisin. *Appl. Microbiol. Biotechnol.* **2010**, *87*, 729–737. [CrossRef] [PubMed]

26. Nostro, A.; Scaffaro, R.; D'Arrigo, M.; Botta, L.; Filocamo, A.; Marino, A.; Bisignano, G. Study on carvacrol and cinnamaldehyde polymeric films: Mechanical properties, release kinetics and antibacterial and antibiofilm activities. *Appl. Microbiol. Biotechnol.* **2012**, *96*, 1029–1038. [CrossRef] [PubMed]

27. Nostro, A.; Scaffaro, R.; D'Arrigo, M.; Botta, L.; Filocamo, A.; Marino, A.; Bisignano, G. Development and characterization of essential oil component-based polymer films: A potential approach to reduce bacterial biofilm. *Appl. Microbiol. Biotechnol.* **2013**, *97*, 9515–9523. [CrossRef] [PubMed]

28. Nostro, A.; Scaffaro, R.; Botta, L.; Filocamo, A.; Marino, A.; Bisignano, G. Effect of temperature on the release of carvacrol and cinnamaldehyde incorporated into polymeric systems to control growth and biofilms of *Escherichia coli* and *Staphylococcus aureus*. *Biofouling* **2015**, *31*, 639–649. [CrossRef] [PubMed]

29. Perale, G.; Casalini, T.; Barri, V.; Müller, M.; Maccagnan, S.; Masi, M. Lidocaine release from polycaprolactone threads. *J. Appl. Polym. Sci.* **2010**, *117*, 3610–3614. [CrossRef]

30. Scaffaro, R.; Botta, L.; Marineo, S.; Puglia, A.M. Incorporation of nisin in poly (ethylene-co-vinyl acetate) films by melt processing: A study on the antimicrobial properties. *J. Food Prot.* **2011**, *74*, 1137–1143. [CrossRef] [PubMed]

31. Scaffaro, R.; Botta, L.; Gallo, G. Photo-oxidative degradation of poly(ethylene-co-vinyl acetate)/nisin antimicrobial films. *Polym. Degrad. Stab.* **2012**, *97*, 653–660. [CrossRef]

32. Campos-Requena, V.H.; Rivas, B.L.; Pérez, M.A.; Garrido-Miranda, K.A.; Pereira, E.D. Polymer/clay nanocomposite films as active packaging material: Modeling of antimicrobial release. *Eur. Polym. J.* **2015**, *71*, 461–475. [CrossRef]

33. Caço, A.I.; Varanda, F.; Pratas de Melo, M.J.; Dias, A.M.A.; Dohrn, R.; Marrucho, I.M. Solubility of Antibiotics in Different Solvents. Part II. Non-Hydrochloride Forms of Tetracycline and Ciprofloxacin. *Ind. Eng. Chem. Res.* **2008**, *47*, 8083–8089. [CrossRef]

34. Botta, L.; Scaffaro, R.; La Mantia, F.P.; Dintcheva, N.T. Effect of different matrices and nanofillers on the rheological behavior of polymer-clay nanocomposites. *J. Polym. Sci. Part B Polym. Phys.* **2010**, *48*, 344–355. [CrossRef]

35. Di Maio, L.; Garofalo, E.; Scarfato, P.; Incarnato, L. Effect of polymer/organoclay composition on morphology and rheological properties of polylactide nanocomposites. *Polym. Compos.* **2015**, *36*, 1135–1144. [CrossRef]

36. Narimissa, E.; Gupta, R.K.; Kao, N.; Choi, H.J.; Jollands, M.; Bhattacharya, S.N. Melt rheological investigation of polylactide-nanographite platelets biopolymer composites. *Polym. Eng. Sci.* **2014**, *54*, 175–188. [CrossRef]

37. Li, Y.; Zhu, J.; Wei, S.; Ryu, J.; Sun, L.; Guo, Z. Poly (propylene)/graphene nanoplatelet nanocomposites: melt rheological behavior and thermal, electrical, and electronic properties. *Macromol. Chem. Phys.* **2011**, *212*, 1951–1959. [CrossRef]

38. Saha, N.R.; Sarkar, G.; Roy, I.; Rana, D.; Bhattacharyya, A.; Adhikari, A.; Mukhopadhyay, A.; Chattopadhyay, D. Studies on methylcellulose/pectin/montmorillonite nanocomposite films and their application possibilities. *Carbohydr. Polym.* **2016**, *136*, 1218–1227. [CrossRef] [PubMed]

39. Keawchaoon, L.; Yoksan, R. Preparation, characterization and *in vitro* release study of carvacrol-loaded chitosan nanoparticles. *Coll. Surf. B Biointerfaces* **2011**, *84*, 163–171. [CrossRef] [PubMed]

MDPI

Article

Silica-Gentamicin Nanohybrids: Synthesis and Antimicrobial Action

Dina Ahmed Mosselhy [1,2,*], Yanling Ge [1], Michael Gasik [1], Katrina Nordström [3], Olli Natri [3] and Simo-Pekka Hannula [1]

[1]   Department of Materials Science and Engineering, School of Chemical Technology, Aalto University, 02150 Espoo, Finland; yanling.ge@aalto.fi (Y.G.); michael.gasik@aalto.fi (M.G.); simo-pekka.hannula@aalto.fi (S.-P.H.)
[2]   Microbiological Unit, Fish Diseases Department, Animal Health Research Institute, Dokki, 12618 Giza, Egypt
[3]   Department of Biotechnology and Chemical Technology, School of Chemical Technology, Aalto University, 02150 Espoo, Finland; katrina.nordstrom@aalto.fi (K.N.); olli.natri@aalto.fi (O.N.)
*   Correspondence: dina.mosselhy@aalto.fi; Tel.: +358-50-408-3533

Academic Editor: Mauro Pollini
Received: 12 December 2015; Accepted: 29 February 2016; Published: 5 March 2016

Abstract: Orthopedic applications commonly require the administration of systemic antibiotics. Gentamicin is one of the most commonly used aminoglycosides in the treatment and prophylaxis of infections associated with orthopedic applications, but gentamicin has a short half-life. However, silica nanoparticles (SiO_2 NPs) can be used as elegant carriers for antibiotics to prolong their release. Our goal is the preparation and characterization of SiO_2-gentamicin nanohybrids for their potential antimicrobial administration in orthopedic applications. *In vitro* gentamicin release profile from the nanohybrids (gentamicin-conjugated SiO_2 NPs) prepared by the base-catalyzed precipitation exhibited fast release (21.4%) during the first 24 h and further extension with 43.9% release during the five-day experiment. Antimicrobial studies of the SiO_2-gentamicin nanohybrids *versus* native SiO_2 NPs and free gentamicin were performed against *Bacillus subtilis* (*B. subtilis*), *Pseudomonas fluorescens* (*P. fluorescens*) and *Escherichia coli* (*E. coli*). SiO_2-gentamicin nanohybrids were most effective against *B. subtilis*. SiO_2 NPs play no antimicrobial role. Parallel antimicrobial studies for the filter-sterilized gentamicin were performed to assess the effect of ultraviolet (UV)-irradiation on gentamicin. In summary, the initial fast gentamicin release fits the need for high concentration of antibiotics after orthopedic surgical interventions. Moreover, the extended release justifies the promising antimicrobial administration of the nanohybrids in bone applications.

Keywords: silica nanoparticles; gentamicin; *in vitro* release; antimicrobial effect; orthopedic applications

1. Introduction

Orthopedic applications such as bone implants and open fractures necessitate therapeutic and prophylactic administration of antibiotics [1]. Microorganisms can attach onto the nails used for stabilization of fractured-bones and may lead to systemic antibiotic resistant biofilms [2]. The traditional systemic administration of antibiotics shows poor penetration to the infected tissues [3]. However, delivery systems that are able to locally release the antibiotics can maintain antibiotic concentrations at the infected sites until healing is complete [4], reduce the frequency of drug administration [5,6] and, consequently, enhance patient compliance [7]. Materials implemented as carriers for antibiotics should have release kinetics that comply with the requirements to treat the infection [8] and should be biodegradable to exclude further surgical intervention to remove them [3,9].

Gram-negative bacteria are increasingly associated with the risk of infections leading to osteomyelitis, which has been more traditionally attributed to gram-positive bacteria, especially *Staphylococcus* species. This increasing risk is due to the high number of administration of orthopedic implants. However, there are inadequate data regarding the involvement of gram-negative bacteria in osteomyelitis [10]. Gentamicin is an aminoglycoside that is ideally used in treatment of osteomyelitis. A potent broad-spectrum antibiotic that is effective against gram-positive and gram-negative bacteria [11]. Controlled release of aminoglycosides is desirable, as their half-life is short and their bioavailability is low, which is a challenge for the conventional treatment modalities [12].

In recent decades, much attention has been paid to silica nanoparticles (SiO_2 NPs) as promising carriers for controlled drug delivery [13,14]. Studies have been performed to investigate the role of SiO_2 xerogels as carriers to enable the controlled release of drugs. Aughenbaugh *et al.* [8] studied the administration of SiO_2 NPs prepared using different water/alkoxysilane molar ratios (4, 6 and 10) as carriers for vancomycin and detected that the high molar ratio SiO_2 gels possessed faster vancomycin release than the low molar ratio ones. Shi *et al.* [15] examined the *in vitro* sustained release of gentamicin from poly(lactide-*co*-glycolide)/mesoporous SiO_2-hydroxyapatite composite material (PLGA/HMS-HA). The release rate of gentamicin from HMS-HA particles was extended to one month suggesting that PLGA/HMS-HA scaffolds are promising drug delivery materials for orthopedic applications. Xue and Shi [6] suggested also that PLGA/mesoporous SiO_2 hybrid structure is an ideal drug release material for bone filling applications. Gentamicin exhibited a sharp initial burst for one day followed by a slow release from the loaded SiO_2 for three weeks. By the encapsulation with PLGA, the release period can be extended to five weeks.

The aforementioned studies have only demonstrated the key role played by SiO_2 NPs as carriers for antibiotics especially in bone applications. Moreover, none of the previous studies have dealt with the antimicrobial properties of the antibiotic-loaded SiO_2 NPs that enable their administration in different orthopedic applications. Accordingly, in the present paper, we report the antimicrobial performance of the synthesized SiO_2-gentamicin nanohybrids. To the best of our knowledge, only a few studies have been implemented regarding the antimicrobial action of gentamicin-conjugated SiO_2 NPs. Corrêa *et al.* [16] detected that SiO_2-encapsulated gentamicin prepared by precipitation route and thus having a high amount of gentamicin on the surface of SiO_2 NPs exhibited effective antimicrobial action against *Staphylococcus aureus* (*S. aureus*). However, the previous work was not coupled with mechanistic studies for the gentamicin release nor for the bacterial interactions with the gentamicin-conjugated SiO_2 NPs. Agnihotri and coworkers [13] examined the antimicrobial action of aminoglycoside-conjugated SiO_2 NPs and detected improved antimicrobial properties of the conjugate against kanamycin-resistant *E. coli*. However, as in other prior studies, the aminoglycoside release studies were not run concurrently with the antimicrobial studies. Therefore, the role of SiO_2 NPs as prime antibiotic carriers that permit the extended release of antibiotics to be implemented in orthopedic applications was not demonstrated. The present study aims to demonstrate the controlled release of gentamicin from the SiO_2-gentamicin nanohybrids through the *in vitro* release studies. The present work makes an original contribution by performing the antimicrobial studies concurrently with the *in vitro* release studies in order to estimate the amount of gentamicin released from the SiO_2-gentamicin nanohybrids. Consequently, appropriate concentrations of the nanohybrids that will supply therapeutic levels of the gentamicin can be applied to avoid the possible sub-inhibitory released concentrations of the antibiotic. Furthermore, our work is designed to shed light on the antimicrobial actions facilitated by the SiO_2-gentamicin nanohybrids in comparison with free gentamicin and native SiO_2 NPs to exclude any antimicrobial role that could be played by the SiO_2 NPs and to demonstrate that the antimicrobial effect can be attributed to the antibiotic.

In the present study, we examine the antimicrobial action of the tested materials against different bacterial species and cover the lack of attention that the gram-negative bacteria have received regarding their role in osteomyelitis. This lack of concern was addressed by performing antimicrobial studies against different laboratory strains of *E. coli* and *P. fluorescens* as models of the gram-negative bacteria that are reported in causing osteomyelitis. As our study does not focus on the clinical etiology of osteomyelitis, nor are we looking to identify causative agents of this condition, we selected *B. subtilis* as a model organism for gram-positive bacteria, as studies with *S. aureus* would require working in

Biosafety level-II facilities. Moreover, as the present study primarily focuses on shedding insights on differences between the antimicrobial sensitivities of gram-negative and gram-positive bacteria in a setting where the antibiotic is delivered in the presence of the SiO_2-gentamicin nanohybrids, *B. subtilis* was chosen as a representative of gram-positive bacteria. Moreover, in this paper we strive to complement previously published work by evaluating the effect of sterilization on the loaded-gentamicin before the antimicrobial application through performing parallel antimicrobial studies for the filter-sterilized gentamicin in comparison to the UV-irradiated gentamicin.

The conventional high processing temperatures of SiO_2 NPs may affect the organic molecules, but this can be overcome as sol-gel technology can be accomplished at room temperature [5]. Therefore, this work is targeted towards the preparation of the SiO_2-gentamicin nanohybrids and native SiO_2 NPs at ambient temperature through a single step sol-gel procedure. The present study focuses on the antimicrobial action of the SiO_2-gentamicin nanohybrids against gram-positive and gram-negative bacteria. The specific focus is on the elucidation of the possible use of the SiO_2-gentamicin nanohybrids in the treatment and prophylaxis of bone infections and orthopedic applications. Moreover, Transmission electron microscopy (TEM) studies were performed to detect the antibacterial mechanistic action of the SiO_2-gentamicin nanohybrids on the microscopic level. The interaction of the native SiO_2 NPs and free gentamicin with the bacterial cells was also studied. The main objective of this study is preparation of SiO_2-gentamicin nanohybrids to demonstrate that the released gentamicin has the extended antimicrobial action and could therefore be used efficiently for targeting infectious microorganisms.

2. Results and Discussion

2.1. Characterization of the SiO₂-Gentamicin Nanohybrids and Native SiO₂ NPs

The SiO_2-gentamicin nanohybrids prepared by the base-catalyzed precipitation route were characterized by using different series of instrumental characteristic techniques to determine the different properties of the SiO_2-gentamicin nanohybrids and native SiO_2 NPs. By X-ray diffraction (XRD), the native SiO_2 NPs and SiO_2-gentamicin nanohybrids showed nearly the same XRD pattern of the broad peak with Bragg angle at 2θ around $24°$ (Supplementary Figure S1), which indicates that the SiO_2 NPs used in the present study were amorphous in nature. This amorphous pattern has been reported in previous studies [17].

The Fourier transform infrared (FTIR) spectra of the native SiO_2 NPs, free gentamicin and SiO_2-gentamicin nanohybrids are depicted in Figure 1A,B. The native SiO_2 NPs (Figure 1A) showed Si–O–Si bending at 553, 790 and 955 cm^{-1}. In addition, a sharp peak was detected at 1053 cm^{-1} for the asymmetric Si–O–Si stretch and a broad band at 3381 cm^{-1} associated with the Si–OH stretching. Similar spectrum was reported previously [18] for the well polymerized SiO_2 network. The free gentamicin (Figure 1A) showed a band at 606 cm^{-1} that is considered a major band for gentamicin [19]. Two more bands at 1524 and 1614 cm^{-1} were also detected. These bands can be ascribed to the N–H vibrational bending of primary aromatic amines [20]. The spectrum of SiO_2-gentamicin nanohybrids (Figure 1B) showed peaks at 559, 793, and 957 and a broad peak at 3199 cm^{-1}. This spectrum did not impose significant shifts from the bands of the native SiO_2 NPs, which can be attributed to the amorphous nature of the SiO_2 NPs. The intensity of the bands decreased after the conjugation. The present measurements showed in addition a shoulder at 606 cm^{-1} and a band at 1529 cm^{-1} that are obviously originating from the gentamicin. The current results indicate that gentamicin has been loaded to the SiO_2 NPs and causes no significant changes in the silica xerogel network.

The scanning electron microscopy (SEM) images provided important information for the native SiO_2 NPs and the SiO_2-gentamicin nanohybrids as shown in Figure 2. The native SiO_2 NPs (Figure 2A) presented smooth spherical surfaces. However, the SiO_2-gentamicin nanohybrids (Figure 2B) showed surface roughness due to the conjugation of gentamicin to the SiO_2 NPs and at the same time, some nanohybrids are coalesced into larger agglomerates. Our results showed the loading of gentamicin to the SiO_2 NPs. The detected roughness of the SiO_2-gentamicin nanohybrids is important, since surface roughness plays a major role in controlling the initial release of antibiotics, as rough surfaces establish a larger area for antibiotic release [21]. Our results suggest that the initial antibiotic release is mainly a surface phenomenon.

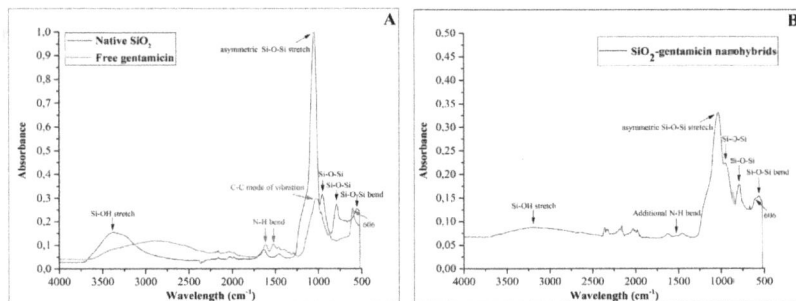

Figure 1. FTIR spectra of the: (**A**) native SiO_2 NPs and free gentamicin sulfate; and (**B**) SiO_2-gentamicin nanohybrids.

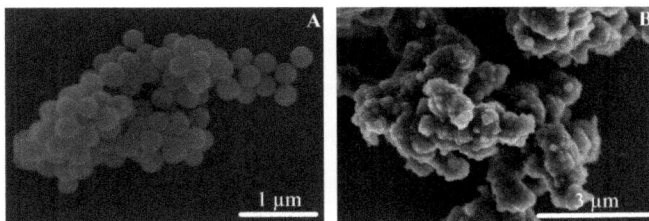

Figure 2. SEM images of the: (**A**) native SiO_2 NPs with the smooth spherical surfaces; and (**B**) SiO_2-gentamicin nanohybrids showing the surface roughness that is caused by the loaded gentamicin.

As depicted in Figure 3, the transmission electron microscopy (TEM) image of the native SiO_2 NPs (Figure 3A) showed a homogenous spherical morphology and an average size of 327 ± 10 nm, with a size range (Figure 3C) of 308 to 339 nm. The SiO_2-gentamicin nanohybrids (Figure 3B) showed network of conglomerated spherical SiO_2 NPs of 332 ± 15 nm average size and a size distribution (Figure 3D) of 292 to 354 nm. The slight increase in the size of SiO_2-gentamicin nanohybrids over that of the native SiO_2 NPs can be related to the loading of the antibiotic to the SiO_2 network.

The mass weight of the overnight dried SiO_2 NPs and SiO_2-gentamicin nanohybrids was measured and their weight difference can be attributed to the loaded gentamicin sulfate that theoretically comprises 12.7 wt % payload. The thermogravimetric analysis (TGA) results of the native SiO_2 NPs (not loaded with the antibiotic) and SiO_2-gentamicin nanohybrids are shown in Figure 4. The data demonstrate a total mass loss of 12.96% and 20.84% upon heating up to 500 °C in an argon atmosphere for the native SiO_2 NPs and SiO_2-gentamicin nanohybrids, respectively. An initial weight loss of 6.94% and 3.18% during heating of the native SiO_2 NPs and SiO_2-gentamicin nanohybrids up to 100 °C, respectively, can be ascribed to the removal of the absorbed and residual water. Hence, the mass loss of the native SiO_2 after heating to 500 °C with the subtracted low-temperature moisture removal is ~6%. The SiO_2-gentamicin nanohybrids showed further weight loss of 4.4% at 215 °C that can be considered as the start of aminoglycoside decomposition. A prior study [22] has recorded the weight loss of amikacin sulfate at the range of 190 to 270 °C. Other work [23] demonstrated that rising the temperature above 150 to 200 °C resulted in the loss of gentamicin. In the present study, the final weight loss of 13.25% in the range of 220 to 500 °C can be related to the removal of glycosidic moieties [13]. The amount of gentamicin in the SiO_2-gentamicin nanohybrids can be determined by subtracting the mass loss of native SiO_2 NPs from the mass loss of SiO_2-gentamicin nanohybrids after excluding the moisture weight loss in both samples and assuming that SiO_2 absorbs equivalent amount of water. Therefore, gentamicin sulfate, according to the TGA data, constitutes ~11.7 wt % of the SiO_2-gentamicin nanohybrids. This is very close to our calculated theoretical loaded gentamicin sulfate (12.7 wt %) to the SiO_2-gentamicin nanohybrids.

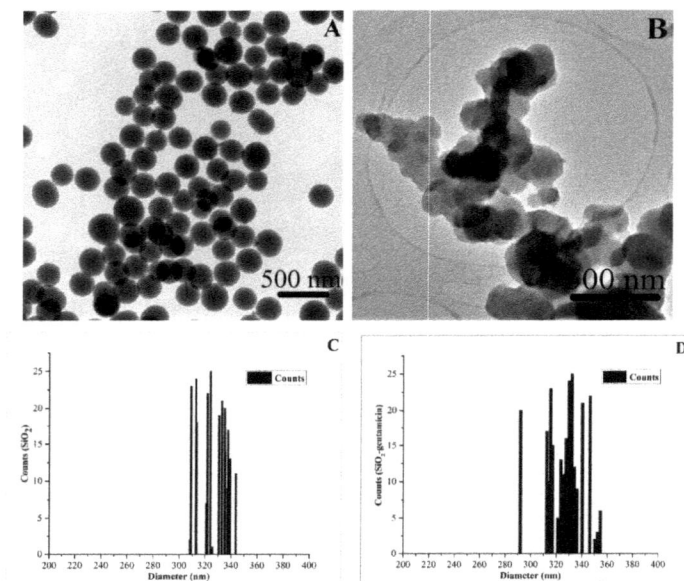

Figure 3. TEM images of the: (**A**) native SiO_2 NPs showing a homogenous spherical morphology; and (**B**) SiO_2-gentamicin nanohybrids displaying the conjugated SiO_2 network in conglomerates. Size distributions of the: (**C**) native SiO_2 NPs; and (**D**) SiO_2-gentamicin nanohybrids. The size distribution was determined by measuring the area of each single NP, using the ImageJ software and calculating the average diameter of the measured 25 NPs.

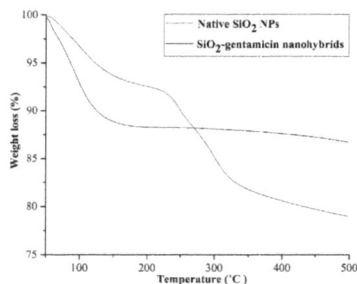

Figure 4. TGA of the native SiO_2 NPs and SiO_2-gentamicin nanohybrids. The mass loss from 215 to 500 °C in the SiO_2-gentamicin nanohybrids was ascribed to gentamicin decomposition.

2.2. *The* in Vitro *Release Profile of Gentamicin from the SiO₂-Gentamicin Nanohybrids*

Another objective of our study, after the detailed characterization of the prepared materials is to detect the sustained gentamicin release from the SiO_2-gentamicin nanohybrids. The *in vitro* release of gentamicin is depicted in Figure 5. The SiO_2-gentamicin nanohybrids (10 mg/mL) containing 11.7 wt % gentamicin sulfate demonstrated a cumulative release of 250.5 ± 1.6, 356 ± 29.9, 386 ± 10.3, 435.3 ± 7.8 and 512.8 ± 0.7 µg/mL gentamicin in PBS solution after 24, 48, 72, 96 and 120 h, respectively. This release profile constitutes a fractional release around 21.4%, 30.5%, 33%, 37.2% and 43.9% of the total amount of the loaded-gentamicin sulfate, correspondingly, over the 5 days of the experiment.

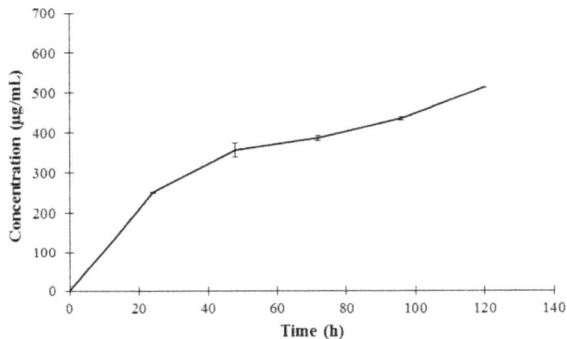

Figure 5. The *in vitro* release of gentamicin from the SiO_2-gentamicin nanohybrids (10 mg/mL) with a drug concentration of 11.7 wt % in PBS at pH 7.4. The mean cumulative concentration of gentamicin (µg/mL) (each data point is the mean of three measurements) was expressed as a function of immersion time (h). The data were linearly fitted to the gentamicin standards. Error bars represent the standard errors ($n = 3$).

The relatively faster gentamicin release during the first 24 h (21.4%) when compared with the slower sustained release during the successive days of the experiment is likely resulting from the initial easily-released gentamicin that is attached to the surface of SiO_2 NPs. Shi *et al.* [15] recorded even higher initial bursts of gentamicin release (60%) from HMS-HA during the first hour. However, slower release was detected within the following 11 h. They proposed that this initial burst release may have also been due to the release of gentamicin molecules that were adsorbed either on the surface of particles or by physical interaction with HMS-HA.

The subsequent decrease in the gentamicin release rate that is observed in our experiment can be attributed to the chemical equilibrium between the gentamicin-silica chemical bonding and the dissolution of water-soluble gentamicin in the aqueous medium [6]. In the same vein, Kortesuo *et al.* [14] showed slower release of the loaded toremifene citrate and dexmedetomidine HCl from the silica gel microspheres after their primary high release due to the depletion of the drugs from the SiO_2 surface.

At fitting our release data to various mathematical models, the data (Table 1) showed the best pharmacokinetic fit to Korsmeyer–Peppas model (Figure 6A) with a release rate constant (k) = 0.966, n = 0.416 ($n \leqslant 0.5$ in case of Fickian diffusion) and a good linearity (R^2 value) of 0.9769. This fit suggests that gentamicin release from the SiO_2-gentamicin nanohybrids is controlled by Fickian diffusion [20,22]. Nevertheless, the release data fit fairly well to first-order model (Figure 6B) with R^2 value of 0.9655; this can explain the initial fast release of gentamicin during the first 24 h. Previous studies have reported that fitting quality is likely better for the initial stage than for the terminal one [24]. We conclude that, gentamicin release is fast at the initial stage, but without an unfavorable initial burst release, followed by slower release during the successive days of the experiment. This release mechanism is due to the amount of gentamicin molecules adsorbed on the surface of the SiO_2 NPs combined with the gentamicin diffusion from the SiO_2-gentamicin nanohybrids.

Table 1. Calculated release rate constants (K_s) and correlation coefficients (R^2) after fitting gentamicin release profile, expressed by various mathematical models.

Mathematical Models	K_s	R^2
Zero-order model	2.5167	0.9662
First-order model	1.1519	0.9655
Higuchi model	0.8583	0.9721
Korsmeyer–Peppas model	0.966	0. 9769

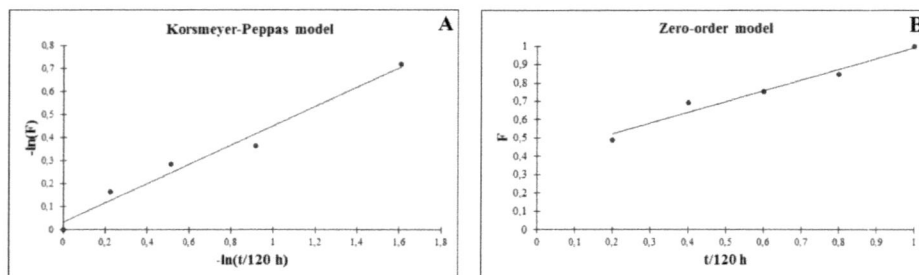

Figure 6. Pharmacokinetic fit of the release profile of gentamicin from the SiO_2-gentamicin nanohybrids to: (**A**) Korsmeyer–Peppas model; and (**B**) zero-order model. F is the drug release fraction at time t.

Our findings suggest that the SiO_2 NPs can be proposed as elegant carriers for gentamicin sulfate that permit the initial fast gentamicin release, which exceeds the minimal gentamicin therapeutic concentration (1 to 4 µg/mL) against different sensitive bacteria [25]. The present findings thus fit the initial need for high concentration of antibiotic after orthopedic surgery and avoid the sub-inhibitory antibiotic concentrations that can provoke a problematic bacterial resistance. In addition, the reported extended release of gentamicin is crucial factor for the efficient implementation of the SiO_2-gentamicin nanohybrids in versatile orthopedic applications.

2.3. Antimicrobial Activity of the SiO_2-Gentamicin Nanohybrids, SiO_2 NPs and Free Gentamicin

2.3.1. Agar Diffusion Assay

The final objective of our study is to identify the antimicrobial properties of the SiO_2-gentamicin nanohybrids that could permit their administration in different orthopedic applications. In order to assess this antimicrobial activity, the zones of bacterial growth inhibition were measured for the conjugated and native SiO_2 NPs in comparison with that of free gentamicin (Figure 7; Supplementary Figure S2). The native SiO_2 NPs were not antimicrobial, as no zones of inhibition were produced against any of the bacterial species that were tested. Our results indicate that the SiO_2 NPs have no antimicrobial effect when they are conjugated with antibiotics and act only as carriers for the gentamicin. Therefore, the antimicrobial effect of the SiO_2-gentamicin nanohybrids is exerted only by the released gentamicin from the SiO_2-gentamicin nanohybrids.

The agar diffusion results showed that the SiO_2-gentamicin nanohybrids and free gentamicin have a more marked antimicrobial effect against the gram-positive bacterium, *B. subtilis*, than on the gram-negative bacteria, *P. fluorescens* and *E. coli*. It is evident that the outer membrane of the gram-negative bacteria most probably acts as an extra barrier to the antimicrobial effects of the gentamicin. Gram-positive bacteria lack this outer membrane that may explain their greater sensitivity [16].

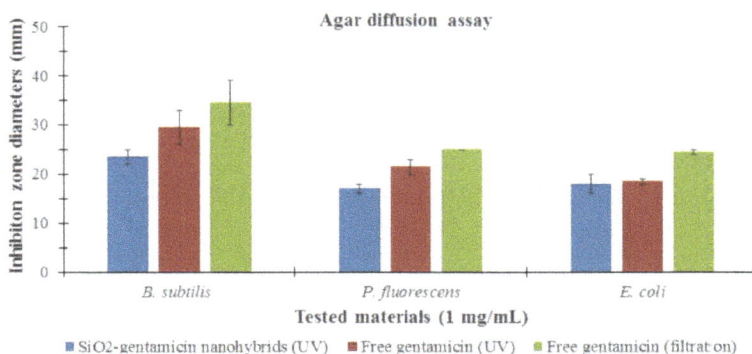

Figure 7. Inhibition zone diameters (mm) of the SiO_2-gentamicin nanohybrids and free gentamicin (sterilized by UV-irradiation and filtration). Error bars represent the standard errors.

2.3.2. MIC (Minimum Inhibitory Concentration) of the SiO_2-Gentamicin Nanohybrids, Native SiO_2 NPs and Free Gentamicin

MIC of the tested twofold serially diluted SiO_2-gentamicin nanohybrids, native SiO_2 NPs and free gentamicin were determined and are shown in Table 2. The native SiO_2 NPs did not inhibit the bacterial growth even at the highest concentration (1 mg/mL). These results confirm our negative agar diffusion assay results for the native SiO_2 NPs and demonstrate that the SiO_2 NPs have no antimicrobial effect. The MIC of the SiO_2-gentamicin nanohybrids was 250 µg/mL (that released 6.26 µg/mL gentamicin) against all the bacterial species. When comparing the same tested concentrations of the free gentamicin with that of the SiO_2-gentamicin nanohybrids, we found that free gentamicin concentration of 0.98 µg/mL inhibited *B. subtilis* and 15.63 µg/mL was effective against *P. fluorescens* and *E. coli*. MIC of the free gentamicin against *P. fluorescens* and *E. coli* is within the range of 3.91 to 15.63 µg/mL that is estimated to be 6.26 µg/mL, which is the amount of released gentamicin from the 250 µg/mL SiO_2-gentamicin nanohybrids that inhibited the *P. fluorescens* and *E. coli*. Comparing the same concentrations of the tested SiO_2-gentamicin nanohybrids with that of the free gentamicin is important to prove that the high MIC of the SiO_2-gentamicin nanohybrids (250 µg/mL) against the bacterial species is governed by the amount of released gentamicin (6.26 µg/mL) and not from the gentamicin itself.

Table 2. MIC of the SiO_2-gentamicin nanohybrids, native SiO_2 NPs and free gentamicin.

Tested Materials	MIC (µg/mL)		
	B. subtilis	*P. fluorescens*	*E. coli*
Native SiO_2 NPs	-	-	-
SiO_2-gentamicin nanohybrids	250 *	250 *	250 *
Free gentamicin	0.98	3.91 to 15.63 [¶]	3.91 to 15.63 [¶]

Notes: (-) indicates that the tested material did not inhibit the bacterial growth. (*) represents that this concentration of the SiO_2-gentamicin nanohybrids released 6.26 µg/mL gentamicin that inhibited the bacterial growth. ([¶]) indicates that the MIC of the free gentamicin is within this range (3.91 to 15.63) as the tested concentrations were widely separated.

The agar diffusion assay and MIC results show that the antimicrobial efficacy of the SiO_2-gentamicin nanohybrids is less than that of the free gentamicin. This may be due to the aforementioned release profile of gentamicin from the SiO_2-gentamicin nanohybrids. After 24 h, the concentrations of 1 mg/mL and 250 µg/mL of the SiO_2-gentamicin nanohybrids released 25.05 and 6.26 µg/mL gentamicin, respectively. The same concentrations of 1 mg/mL and 250 µg/mL of the free gentamicin are solely gentamicin sulfate, which explains the more potent antimicrobial efficacy of the free gentamicin than that of the SiO_2-gentamicin nanohybrids. Our data demonstrate that the sustained release of gentamicin from the SiO_2-gentamicin nanohybrids fits their ideal antimicrobial administration in different orthopedic applications. However, careful considerations should be taken regarding the concentrations of the administered antibiotic-conjugated nanoparticles to avoid the release of sub-inhibitory concentrations of the antibiotic. These sub-inhibitory concentrations may result in the generation of antibiotic bacterial resistance that is elicited mainly from the miss-use of the antibiotic-conjugated nanoparticles that would continuously supply low levels of antibiotic. A previous study [23] has demonstrated a higher resistance of *S. aureus* to gentamicin-conjugated Au NPs (MICs were 220 and 440 µg/mL) than to free gentamicin (MIC was 2 µg/mL) and concluded that this resistance is not elicited from the low efficacy of gentamicin itself, but from the released gentamicin that is interacting with the bacterial cells. Accordingly, we stress that it is important to couple *in vitro* antibiotic release experiments with the antimicrobial assays to exclude any bacterial resistance related to the low concentration of the released antibiotic from the carriers and not from the antibiotic itself. According to our results, 10 mg/mL of the SiO_2-gentamicin nanohybrids released 250.5 ± 1.6, 356 ± 29.9, 386 ± 10.3, 435.3 ± 7.8 and 512.8 ± 0.7 µg/mL gentamicin after 24, 48, 72, 96 and 120 h, respectively. In this case, the released gentamicin concentration exceeded the MIC of the free gentamicin every 24 h; however, there was a high standard deviation value for the released gentamicin after 48 h. It is therefore evident that attention should be taken in the formulation of these materials for orthopedic applications. Notably, attention should be paid, on the one hand, to the concentrations of the applied materials to avoid the release of sub-inhibitory concentrations of the antibiotic and on the other hand, to avoid possible toxicity, which may result when the aim is to provide adequately high levels of the released antibiotic. More importantly, the parallel study of the antimicrobial properties of the filter-sterilized gentamicin showed non-significant increase in the diameters of the inhibition zones (Figure 7) in comparison to that of the UV-sterilized free gentamicin. The recorded MIC was exactly the same. These experiments show that UV-irradiation can be used as a convenient method for sterilization of the antibiotic-loaded nanoparticles, especially if the size of the loaded nanoparticles is larger than the pore size of the syringe filter (0.2 µm). Similarly, prior literature [26] described that irradiation did not affect the release of gentamicin from the gentamicin-loaded collagen/PLGA microparticle composite.

2.3.3. Bactericidal Mechanistic Action of the SiO_2-Gentamicin Nanohybrids *versus* Free Gentamicin

In order to provide new insights on the antimicrobial properties of the SiO_2-gentamicin nanohybrids on the microscopic level, TEM studies were performed. The TEM micrographs of the bacterial cells treated with the native SiO_2 NPs, SiO_2-gentamicin nanohybrids and free gentamicin, as well as the untreated cells are depicted in Figure 8. The untreated *B. subtilis* (Figure 8A) showed the typical thick cell wall of gram-positive bacteria, while the untreated *P. fluorescens* (Figure 8B) demonstrated the thin cell wall of gram-negative bacteria. The cell wall thickness of gram-positive bacterium *B. subtilis* is about 20–30 nm and formed by interwoven peptidoglycans and secondary polymers. The gram-negative bacteria possess thin peptidoglycan layer with an outer lipopolysaccharide membrane [27].

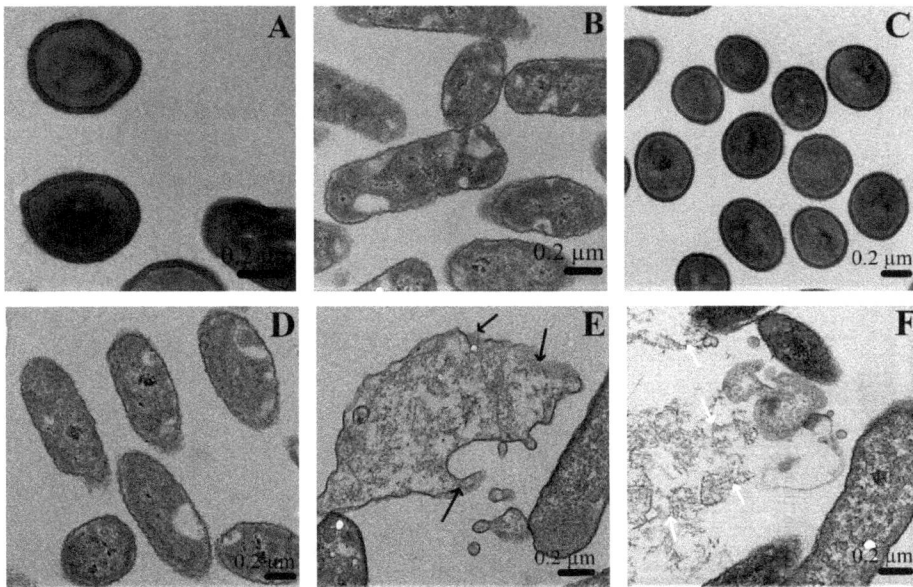

Figure 8. TEM micrographs of the: (**A**) untreated *B. subtilis* and (**B**) *P. fluorescens* (control negative); native SiO$_2$ NPs-treated (**C**) *B. subtilis* and (**D**) *P. fluorescens*; and *P. fluorescens* interacted with (**E**) SiO$_2$-gentamicin nanohybrids and (**F**) free gentamicin. Black arrows indicate disorganization of the bacterial cell membranes. White arrows show complete deterioration of the bacterial cells.

The native SiO$_2$ NPs-treated bacterial cells (Figure 8C,D) showed no morphological changes and the bacterial cells appeared intact and similar to the control untreated cells. These results confirm the lack of any antimicrobial properties for the SiO$_2$ NPs at the microscopic level and show that only the gentamicin is responsible for the antimicrobial effect. The SiO$_2$-gentamicin nanohybrids-treated bacterial cells (Figure 8E) presented disorganization of the bacterial cell membranes. The free gentamicin-treated bacterial cells (Figure 8F) were more affected than the bacterial cells treated with the SiO$_2$-gentamicin nanohybrids. The treatment with the free gentamicin showed complete deterioration of the bacterial cell membranes resulting in the leakage of intracellular contents. The mechanism by which the bactericidal action of the SiO$_2$-gentamicin nanohybrids and free gentamicin is exerted can be attributed to the aminoglycoside structure of gentamicin. The binding sites for aminoglycosides are teichoic acids of the cell wall and phospholipids of the cell membrane of gram-positive bacteria [28]. Amongst Enterobacteriaceae members and *Pseudomonas species*, aminoglycosides can cross the cell walls through porin channels and bind to the 30S subunit of ribosome, resulting in errors in protein synthesis and bacterial death [29].

Finally, it should be noted that, our TEM results represent a third method to confirm the agar diffusion assay and MIC outcomes as follows: (1) the SiO$_2$ NPs act as carriers for the sustained release of gentamicin and have no antimicrobial effect themselves; and (2) The SiO$_2$-gentamicin nanohybrids have a lower antimicrobial effect than the free gentamicin. This correlates with the sustained release profile of gentamicin from the SiO$_2$-gentamicin nanohybrids, which supports their potential to be used in different orthopedic applications that require extended antibiotic release. However, it is to be noted that this study has been only focused on model organisms, and not on bacteria isolated from bone infections and associated with osteomyelitis. It would therefore be paramount to use such organisms in further studies to show more concrete evidence for the potential use of SiO_2-gentamicin nanohybrids for treatment and prophylaxis of osteomyelitis

3. Materials and Methods

3.1. Synthesis and Characterization of the SiO_2-Gentamicin Nanohybrids and Native SiO_2 NPs

The SiO_2-gentamicin xerogels were prepared via the base-catalyzed precipitation route adopting the method described by Corrêa *et al.* [16]. Briefly, 500 mg of gentamicin sulfate (Sigma-Aldrich, Shanghai, China) was dissolved in 10 mL of tetraethyl orthosilicate (TEOS, Aldrich® chemistry, Steinheim, Germany). Then, 20 mL of ammonium hydroxide (28%–30%, Sigma-Aldrich, St. Louis, MO, USA) was added to the solution. The mixture was stirred for 20 min at room temperature until precipitation. The resultant material was dried overnight at room temperature and then ground. The native SiO_2 NPs were prepared with the same aforementioned method but excluding the addition of gentamicin sulfate.

Once the materials were prepared, several instrumental techniques were applied for the characterization. The SiO_2-gentamicin nanohybrids and native SiO_2 powders were subjected to X-ray diffraction measurements with an X′pert Powder Pro (PANalytical/PW3040/60, Almelo, The Netherlands) operated at generator settings of 40 mA and 45 kV using Cu-Kα radiation and a goniometer scanning (2θ) over 10°–90°. The FT-IR measurements were performed using a Nicolet 380 FT-IR (Thermo Electron Corporation, White Bear Lake, MN, USA) spectrometer in the attenuated total reflection (ATR) mode, with a resolution of 2 cm^{-1} and a scan range of 4000–500 cm^{-1}. The spectra were averaged over 64 scans. The samples were analyzed using the absorbance mode. The SEM analyses were accomplished by using a Hitachi SEM (S-4700, Tokyo, Japan) to detect the surface morphology of the synthesized materials. For the SEM studies, the samples of the SiO_2-gentamicin nanohybrids were coated with a thin carbon film using a cool sputtering device (Leica EM SCD050, Wetzlar, Germany). The TEM studies of the shape and size of the prepared materials were implemented using a Tecnai G2 F20 TEM (FEI, Eindhoven, The Netherlands). The mass changes of the native SiO_2 NPs and SiO_2-gentamicin nanohybrids were weighted after overnight drying at room temperature to determine the theoretical concentration of the loaded drug [6,15]. The TGA of the native and conjugated SiO_2 NPs was performed with a simultaneous thermal and spectral analysis using a STA449C Jupiter (Netzsch Gerätebau GmbH, Selb, Germany) coupled with a Tensor-27 FTIR spectrometer (Bruker Optics, Ettlingen, Germany). The analysis was conducted from 30 to 500 °C at a rate of 10 °C/min following Sharma *et al.* [22], but in an argon atmosphere (flow rate of 20 mL/min). Aluminum crucibles without lids were utilized in the experiment, with an empty crucible as a reference.

3.2. The in Vitro Release Studies of Gentamicin from the SiO_2-Gentamicin Nanohybrids

Gentamicin sulfate absorbs neither ultraviolet nor visible light. Hence, indirect spectrophotometric analysis was performed to detect the concentrations of the antibiotic using derivatization. Phthaldialdehyde reagent (Sigma-Aldrich, St. Louis, MO, USA) acted as a derivatizing agent that interacts with the amino groups of gentamicin sulfate to produce a chromophoric product [30]. The *in vitro* release studies were performed in phosphate-buffered saline solution (PBS, pH 7.4) at 37 °C. The SiO_2-gentamicin nanohybrids (20 mg) were placed in tubes containing 2 mL PBS (10 mg/mL). The tubes were incubated at 37 °C and shaken at 150 rpm (Kühner incubator shaker, Lab-Therm, Fennolab, Basel, Switzerland) for regular time intervals of 24, 48, 72, 96 and 120 h. After incubation, the tubes were centrifuged at 5000 rpm (Centrifuge 5424, Eppendorf AG, Hamburg, Germany) for 5 min. Then, PBS medium was discarded and 2 mL of fresh PBS (pH 7.4) was added to remove all the unattached gentamicin sulfate molecules [25]. Two milliliters of isopropyl alcohol (Sigma-Aldrich, Steinheim, Germany) and another 2 mL of phthaldialdehyde reagent were added to the fresh PBS. The gentamicin sulfate complex concentration was detected by an UV-vis spectrophotometer (Hitachi, U-2000, Hachioji City, Tokyo, Japan) at 332 nm [30]. The gentamicin standards were prepared by dissolving appropriate amounts of gentamicin (10, 100 μg/mL and 1 mg/mL) in PBS. The release rate of gentamicin from the SiO$_2$-gentamicin nanohybrids was detected throughout 5 days. Statistical analysis was conducted using the statistical package of Microsoft Excel 2013 (Office Professional Plus 2013, Impressa Systems, Santa Rosa, CA, USA). Each mean value was calculated from 3 consecutive measurements. The standard deviations are represented by the ± values. Confidence interval of mean values was detected as 95%.

Various mathematical models were applied to analyze the pharmacokinetic release profile of the gentamicin from the SiO$_2$-gentamicin nanohybrids. These models are: (1) zero-order model; (2) first-order model; (3) Higuchi square root of time model; and (4) Korsmeyer-Peppas model as shown by Equations (1)–(4), respectively:

$$F = K^* t \tag{1}$$

$$F = 1 - e^{-k^* t} \tag{2}$$

$$F = K^* t^{1/2} \tag{3}$$

$$F = K^* t^n \tag{4}$$

In the equations, F is the drug release fraction at time t ($F = \dfrac{\%Q_t}{\%Q_\infty}$ in which $\%Q_t$ is the drug-released percentage at time t and $\%Q_\infty$ is the total drug-released percentage) and K^* represents the release rate constant. The exponent (0.5, 1 or n) is related to the diffusional processes that could describe the different mechanisms of gentamicin release. The value of the exponent changes according to the release mechanism. In case of Fickian diffusion $n \leqslant 0.5$, for anomalous transport $0.5 < n < 1$ and for zero-order release $n = 1$ [20,22]. Regression analysis was used to detect which model corresponds to our *in vitro* release data.

3.3. Comparison of the Antimicrobial Action of the SiO$_2$-Gentamicin Nanohybrids Versus Native SiO$_2$ NPs and Free Gentamicin

3.3.1. Agar Diffusion Assay

Antimicrobial susceptibility testing is most commonly performed using the agar diffusion method, as recommended by the Clinical and Laboratory Standards Institute (CLSI) [31]. The sensitivity of the bacteria to the tested materials was determined by measuring the inhibition zones produced by diffusion of the antibacterial agents from the wells into the surrounding medium. The results were interpreted according to the procedure presented by Winn *et al.* [32] and Quinn *et al.* [33].

Three different bacterial species from the culture collection of the Department of Biotechnology and Chemical Technology, Aalto University were used as representatives of gram-positive and gram-negative bacteria; namely, *B. subtilis* (TKK 10151), *P. fluorescens* and *E. coli* (VTT E-94564) that were stored in −80 °C. The bacterial species were sub-cultured on Nutrient agar (Lab M Limited®, Heywood, UK). Colonies obtained from the agar plates after 24 h incubation were inoculated into Mueller Hinton broth (Lab M Limited®). The bacterial concentrations were detected by measuring the absorbance using an UV-vis spectrophotometer (Hitachi, U-2000, Tokyo, Japan) at 600 nm [34]. Since the optical density measures the viable and dead bacteria in suspensions and as the traditional criterion to differentiate viable from dead bacterial cells is the ability of viable cells to produce a colony [35], our suspensions were obtained from 24-h cultured plates and the cultured broth was adjusted to match 0.5 McFarland standard. A volume of 100 μL of each bacterial suspension was spread on the Muller Hinton agar plates and 20 μL of the different tested materials was dispensed in the wells of the plates. Incubation of the plates was executed for 24 h at 28 °C for *B. subtilis* and *P. fluorescens*; and at 37 °C for *E. coli*. The agar diffusion assays were performed in duplicate and the inhibition zone diameters (mm) are the average of the measurements. Standard deviations of the inhibition zones (mm) are represented by the ± values.

Sterilization of the powder materials took place through UV-irradiation using an UV Chamber (GS Gene Linker®, Bio-Rad, Hercules, CA, USA). Then, all the UV-irradiated samples were dispersed in sterile de-ionized water in concentrations of 1 mg/mL for the microbiological tests. The agar diffusion assays were performed at a starting concentration of 1 mg/mL of the tested materials to screen for the antimicrobial action of the SiO_2-gentamicin nanohybrids in comparison with the native SiO_2 NPs and the free gentamicin. Then, the concentrations of the tested materials were decreased to detect the MIC in the broth microdilution assay. The tested materials were sonicated (power 234 W, working frequency 47 kHz ± 6%, Bransonic®, 2210E-DTH, Derwood, MD, USA) for 30 min before the antimicrobial applications to obtain homogenous solutions.

3.3.2. Broth Microdilution Assay

MIC is the minimal inhibitory concentration of antimicrobial agent that inhibits bacterial growth [13]. The standard microdilution method was used to determine the MIC of the SiO_2-gentamicin nanohybrids, native SiO_2 NPs and free gentamicin according to the CLSI [31]. The original concentrations of the tested bacterial suspensions were equivalent to 0.5 McFarland standard. Within 15 min of the inoculum preparation, tenfold serial dilutions were performed in Mueller Hinton broth to reach 1.5 to 3×10^5 CFU/mL. For the tested materials, twofold serial dilutions were performed from 1 mg/mL to 0.98 µg/mL. A volume of 100 µL of the different tested materials was added from the concentrations of 1000, 250, 62.5, 15.63, 3.91 and 0.98 µg/mL into the wells of the microdilution plate. Then, 10 µL of the bacterial suspensions were inoculated into the wells. Mueller Hinton broth without any added antimicrobial material was used as a negative control and only bacterial suspensions were used as a positive control. The results of MIC were read after 24 h of incubation.

Parallel antimicrobial tests (Agar diffusion assay and MIC) were performed for the free gentamicin sulfate sterilized by filtration through 25 mm syringe filter w/0.2 µm cellulose acetate membrane (VWR International, Wallkill, NY, USA). These tests were done to detect the possible deteriorating effect of the UV-irradiation on the loaded-gentamicin and whether it can be used as a sterilization option for antibiotic-conjugated nanoparticles or not.

3.3.3. TEM Interaction Studies of the SiO_2-Gentamicin Nanohybrids, Native SiO_2 NPs and Free Gentamicin with the Bacterial Cells

Microscopic interaction of the tested materials with the gram-positive bacterium, *B. subtilis* and the gram-negative bacterium, *P. fluorescens* was investigated by TEM. The study included four groups. The first group consisted of the native bacterial cells that acted as a control negative. The second, third and fourth groups were the bacterial cells treated with the native SiO_2 NPs, SiO_2-gentamicin nanohybrids and free gentamicin, respectively.

Bacterial cells from the 24 h-cultured agar plates were inoculated into Muller Hinton broth and incubated at 28 °C for 24 h at 200 rpm in an incubator shaker (Lab-Therm, Fennolab, Kühner, Basel, Switzerland). 10 µg/mL of the tested materials was added to the bacterial suspensions [36,37] and re-incubation for a further 24 h was performed. The incubated bacterial suspensions with the tested materials (5 mL) were centrifuged at 10,000 rpm for 5 min, washed and re-centrifuged [38]. The samples were fixed by adding 0.5 mL of 2.5% gluteraldehyde at room temperature and collection of cells was accomplished by centrifugation at 13,000 rpm for 3 min. The supernatant was discarded and another 0.5 mL of 2.5% glutaraldehyde was added to the bacterial cells, after which the pellets were kept at 4 °C overnight. Washing of glutaraldehyde was performed using sodium phosphate buffer. Then, 1% osmium tetroxide was used as a second fixative for 1 h at room temperature. Washing of osmium tetroxide was performed two times by sodium phosphate buffer. The dehydration process was accomplished by increasing ethanol concentrations (50%, 70%, 96% and 100%) and finally by acetone (100%). Infiltration of the samples was implemented in ascending series of Epon®:acetone mix, starting with 30% for 3 h, followed by 70% and finally two changes of 100% epon, 3 h for each.

Polymerization of the samples was performed in 60 °C oven for 14 h. After polymerization, the epon blocks were cut into thin sections of 60 nm thicknesses using an ultramicrotome (Leica, EM Ultra Cut UC6ei, Leica Mikrosysteme GmbH, Vienna, Austria). The sections were placed on grids (Formvar carbon film on 300 mesh-Cu grids, Electron Microscopy Science, Hatfield, PA). The sections on the grids were stained with 0.5% uranyl acetate, followed by 3% lead citrate for TEM micrographs.

4. Conclusions

We have demonstrated that the gentamicin of the SiO_2-gentamicin nanohybrids represents 11.7 wt % payload that permits the fast gentamicin release of 21.4% during the first 24 h and extended release of 43.9% over the five-day experiment. This fast release in the initial stage suggests that SiO_2 NPs are promising carriers for antibiotics and may have possibilities in the treatment and prophylaxis of infections associated with orthopedic applications that require high initial concentration of antibiotic for treatment followed by extended release. Coupling the antimicrobial tests of the SiO_2-gentamicin nanohybrids with the *in vitro* release experiments is essential to apply the nanohybrids with concentrations that would ideally release the therapeutic levels of the antibiotic. This could avoid the crucial problem of releasing sub-inhibitory concentrations of the antibiotic. The SiO_2-gentamicin nanohybrids are more potent against the gram-positive *B. subtilis* than the gram-negative *P. fluorescens* and *E. coli* when tested as pure cultures on laboratory media. UV-irradiation is a convenient method for sterilization of the nanoparticles loaded with gentamicin with no adverse effect on the antimicrobial properties of the antibiotic.

This study strives to contribute to the research targeting the facile preparation of SiO_2 NPs to be used as antibiotic carriers that can permit the extended antibiotic release for efficient antimicrobial administration in different orthopedic applications. The present data have shown the different sensitivities of bacteria to the released antibiotic; however, it is evident that more detailed studies on a wider range of bacteria, preferably isolated from patients with osteomyelitis and capable of biofilm formation should be studied. In order to study the full potential of the developed materials, further work on bacterial strains relevant for osteomyelitis, especially the gram-positive *Staphylococcus species* is called for. Moreover, co-culture and bacterial interactions with bone and bone implants would also be of interest. Furthermore, longer *in vitro* antibiotic release studies should be concurrently performed for each specific clinical application of the SiO_2-gentamcin nanohybrids. Our future research is directed towards the *in vitro* and *in vivo* toxicological evaluations of the SiO_2-gentamicin nanohybrids to ascertain their more extensive application potential.

Supplementary Materials: The following are available online at www.mdpi.com/1996-1944/9/3/170/s1. Figure S1: X-ray diffraction (XRD) patterns of the native SiO_2 NPs and SiO_2-gentamicin nanohybrids, Figure S2: Antimicrobial activity of the: (A) native SiO_2 NPs; and (B) SiO_2-gentamicin nanohybrids *versus* free gentamicin against: (1) *B. subtilis*; (2) *P. fluorescens*; and (3) *E. coli*.

Acknowledgments: Dina A. Mosselhy acknowledges the Centre for International Mobility (CIMO) for the funding support (KM-14-9074) of the Finnish government scholarship pool and Aalto University School of Chemical Technology. The authors acknowledge Teemu Myllymäki, Department of Applied Physics, Aalto University for his aid in the FT-IR measurements. We would like to thank Juho Lotta, Department of Materials Science and Engineering, Aalto University for his help in the SEM analyses. We thank Lauri Viitala Department of Chemistry, Aalto University for his fruitful discussions in fitting the gentamicin release data to a suitable model. We are grateful to the EM-Unit, Institute of Biotechnology, University of Helsinki for their assistance in processing the bacteriological samples for TEM studies.

Author Contributions: Dina A. Mosselhy prepared the nanomaterials and performed the antimicrobial tests under the supervision of Simo-Pekka Hannula and Katrina Nordström, respectively. Yanling Ge performed the TEM characterization of the nanomaterials and the nanomaterials-bacterial interactions. Michael Gasik performed the TGA and the statistical data analysis. Olli Natri participated in the design of the antimicrobial experiments. Dina A. Mosselhy wrote the manuscript. All the authors discussed the results, revised and approved the final version of the manuscript.

Conflicts of Interest: The authors declare no conflicts of interest.

References

1. Falaize, S.; Radin, S.; Ducheyne, P. *In Vitro* Behavior of Silica-Based Xerogels Intended as Controlled Release Carriers. *J. Am. Ceram. Soc.* **1999**, *82*, 969–976. [CrossRef]
2. Radin, S.; Ducheyne, P. Controlled Release of Vancomycin from Thin Sol-Gel Films on Titanium Alloy Fracture Plate Material. *Biomaterials* **2007**, *28*, 1721–1729. [CrossRef] [PubMed]
3. Thein, E.; Tafin, U.F.; Betrisey, B.; Trampuz, A.; Borens, O. *In Vitro* Activity of Gentamicin-Loaded Bioabsorbable Beads against Different Microorganisms. *Materials* **2013**, *6*, 3284–3293. [CrossRef]
4. Radin, S.; Ducheyne, P.; Kamplain, T.; Tan, B.H. Silica sol-gel for the controlled release of antibiotics. I. Synthesis, characterization, and *in vitro* release. *J. Biomed. Mater. Res.* **2001**, *57*, 313–320. [CrossRef]
5. Barbé, C.; Bartlett, J.; Kong, L.; Finnie, K.; Lin, H.Q.; Larkin, M.; Calleja, S.; Bush, A.; Calleja, G. Silica Particles: A Novel Drug-Delivery System. *Adv. Mater.* **2004**, *16*, 1959–1966. [CrossRef]
6. Xue, J.M.; Shi, M. PLGA/mesoporous Silica Hybrid Structure for Controlled Drug Release. *J. Control. Release* **2004**, *98*, 209–217. [CrossRef] [PubMed]
7. Gao, P.; Nie, X.; Zou, M.; Shi, Y.; Cheng, G. Recent Advances in Materials for Extended-Release Antibiotic Delivery System. *J. Antibiot.* **2011**, *64*, 625–634. [CrossRef] [PubMed]
8. Aughenbaugh, W.; Radin, S.; Ducheyne, P. Silica Sol-Gel for the Controlled Release of Antibiotics. II. The Effect of Synthesis Parameters on the *In vitro* Release Kinetics of Vancomycin. *J. Biomed. Mater. Res.* **2001**, *57*, 321–326. [CrossRef]
9. Radin, S.; El-Bassyouni, G.; Vresilovic, E.J.; Schepers, E.; Ducheyne, P. *In Vivo* Tissue Response to Resorbable Silica Xerogels as Controlled-Release Materials. *Biomaterials* **2005**, *26*, 1043–1052. [CrossRef] [PubMed]
10. De Carvalho, V.C.; Rosalba, P.; de Oliveira, P.R.D.; Dal-Paz, K.; de Paula, A.P.; Félix, C.D.S.; Lima, A.L.L.M. Gram-Negative Osteomyelitis: Clinical and Microbiological Profile. *Braz. J. Infect. Dis.* **2012**, *16*, 63–67. [CrossRef]
11. Faber, C.; Stallmann, H.P.; Lyaruu, D.M.; Joosten, U.; von Eiff, C.; van Nieuw Amerongen, A.; Wuisman, P.I.J.M. Comparable Efficacies of the Antimicrobial Peptide Human Lactoferrin 1–11 and Gentamicin in a Chronic Methicillin-Resistant *Staphylococcus aureus* Osteomyelitis Model. *Antimicrob. Agents Chemother.* **2005**, *49*, 2438–2444. [CrossRef] [PubMed]
12. Abdelghany, S.M.; Quinn, D.J.; Ingram, R.J.; Gilmore, B.F.; Donnelly, R.F.; Taggart, C.C.; Scott, C.J. Gentamicin-Loaded Nanoparticles Show Improved Antimicrobial Effects towards *Pseudomonas aeruginosa* Infection. *Int. J. Nanomed.* **2012**, *7*, 4053–4063.
13. Agnihotri, S.; Pathak, R.; Jha, D.; Roy, I.; Gautam, H.K.; Sharma, A.K.; Kumar, P. Synthesis and Antimicrobial Activity of Aminoglycoside-Conjugated Silica Nanoparticles Against Clinical and Resistant Bacteria. *New J. Chem.* **2015**, *39*, 6746–6755. [CrossRef]
14. Kortesuo, P.; Ahola, M.; Kangas, M.; Kangasniemi, I.; Yli-Urpo, A.; Kiesvaara, J. *In Vitro* Evaluation of Sol-Gel Processed Spray Dried Silica Gel Microspheres as Carrier in Controlled Drug Delivery. *Int. J. Pharm.* **2000**, *200*, 223–229. [CrossRef]
15. Shi, X.; Wang, Y.; Ren, L.; Zhao, N.; Gong, Y.; Wang, D.A. Novel Mesoporous Silica-Based Antibiotic Releasing Scaffold for Bone Repair. *Acta Biomater.* **2009**, *5*, 1697–1707. [CrossRef] [PubMed]
16. Corrêa, G.G.; Morais, E.C.; Brambilla, R.; Bernardes, A.A.; Radtke, C.; Dezen, D.; Júnior, A.V.; Fronza, N.; Dos Santos, J.H.Z. Effects of the Sol-Gel Route on the Structural Characteristics and Antibacterial Activity of Silica-Encapsulated Gentamicin. *Colloids Surf. B. Biointerfaces* **2014**, *116*, 510–517. [CrossRef] [PubMed]
17. Lee, J.; Lee, Y.; Youn, J.K.; Na, H.B.; Yu, T.; Kim, H.; Lee, S.M.; Koo, Y.M.; Kwak, J.H.; Park, H.G.; *et al.* Simple Synthesis of Functionalized Superparamagnetic Magnetite/silica Core/shell Nanoparticles and Their Application as Magnetically Separable High-Performance Biocatalysts. *Small* **2008**, *4*, 143–152. [CrossRef] [PubMed]
18. Radin, S.; Falaize, S.; Lee, M.H.; Ducheyne, P. *In Vitro* Bioactivity and Degradation Behavior of Silica Xerogels Intended as Controlled Release Materials. *Biomaterials* **2002**, *23*, 3113–3122. [CrossRef]
19. Sarabia-Sainz, A.-I.; Montfort, G.R.C.; Lizardi-Mendoza, J.; Sánchez-Saavedra, M.D.P.; Candia-Plata, M.D.C.; Guzman, R.Z.; Lucero-Acuña, A.; Vazquez-Moreno, L. Formulation and Characterization of Gentamicin-Loaded Albumin Microspheres as a Potential Drug Carrier for the Treatment of *E. coli* K88 Infections. *Int. J. Drug Deliv.* **2012**, *4*, 209–218.

20. Pandey, H.; Parashar, V.; Parashar, R.; Prakash, R.; Ramteke, P.W.; Pandey, A.C. Controlled Drug Release Characteristics and Enhanced Antibacterial Effect of Graphene Nanosheets Containing Gentamicin Sulfate. *Nanoscale* **2011**, *3*, 4104–4108. [CrossRef] [PubMed]

21. Van de Belt, H.; Neut, D.; Uges, D.R.A.; Schenk, W.; van Horn, J.R.; van der Mei, H.C.; Busscher, H.J. Surface Roughness, Porosity and Wettability of Gentamicin-Loaded Bone Cements and Their Antibiotic Release. *Biomaterials* **2000**, *21*, 1981–1987. [CrossRef]

22. Sharma, U.K.; Verma, A.; Prajapati, S.K.; Pandey, H.; Pandey, A.C. *In Vitro, in Vivo* and Pharmacokinetic Assessment of Amikacin Sulphate Laden Polymeric Nanoparticles Meant for Controlled Ocular Drug Delivery. *Appl. Nanosci.* **2014**, *5*, 143–155. [CrossRef]

23. Perni, S.; Prokopovich, P. Continuous Release of Gentamicin from Gold Nanocarriers. *RSC Adv.* **2014**, *4*, 51904–51910. [CrossRef] [PubMed]

24. Radin, S.; Chen, T.; Ducheyne, P. The Controlled Release of Drugs from Emulsified, Sol Gel Processed Silica Microspheres. *Biomaterials* **2009**, *30*, 850–858. [CrossRef] [PubMed]

25. El-Ghannam, A.; Ahmed, K.; Omran, M. Nanoporous Delivery System to Treat Osteomyelitis and Regenerate Bone: Gentamicin Release Kinetics and Bactericidal Effect. *J. Biomed. Res. Part B Appl. Biomater.* **2005**, *73B*, 227–284. [CrossRef] [PubMed]

26. Vetten, M.A.; Yah, C.S.; Singh, T.; Gulumian, M. Challenges Facing Sterilization and Depyrogenation of Nanoparticles: Effects on Structural Stability and Biomedical Applications. *Nanomed. Nanotech. Biol. Med.* **2014**, *10*, 1391–1399. [CrossRef] [PubMed]

27. Beveridge, T.J.; Makin, S.A.; Kadurugamuwa, J.L.; Li, Z. Interactions between biofilms and the environment. *FEMS Microbiol. Rev.* **1997**, *20*, 291–303. [CrossRef] [PubMed]

28. Chen, L.; Zhang, J. Bioconjugated Magnetic Nanoparticles for Rapid Capture of Gram-positive Bacteria. *J. Biosens. Bioelectron.* **2012**, *S11*, 1–5. [CrossRef]

29. Neu, C.H.; Gootz, T.D. Chapter 11: Antimicrobial chemotherapy. In *Medical Microbiology*, 4th ed.; Baron, S., Ed.; University of Texas Medical Branch at Galveston: Galveston, TX, USA, 1996; pp. 1–10.

30. Phromsopha, T.; Baimark, Y. Chitosan Microparticles Prepared by the Water-in-Oil Emulsion Solvent Diffusion Method for Drug Delivery. *Biotechnology* **2010**, *9*, 61–66. [CrossRef]

31. Clinical and Laboratory Standards Institute (CLSI). *Methods for Dilution Antimicrobial Susceptibility Tests for Bacteria that Grow Aerobically*, Approved standard—Ninth Edition; M07-A9; CLSI: Wayre, PA, USA, 2012; Volume 32, pp. 1–68.

32. Winn, W.C., Jr.; Allen, S.; Janda, W.; Koneman, E.W.; Procop, G.; Schreckenberger, P.; Woods, G. *Koneman's Color Atlas and Textbook of Diagnostic Microbiology*, 6th ed.; Koneman, E.W., Ed.; Lippincott Williams & Wilkins: Philadelphia, PA, USA, 2006.

33. Quinn, P.J.; Markey, B.K.; Leonard, F.C.; FitzPatrick, E.S.; Fanning, S.; Hartigan, P.J. *Veterinary Microbiology and Microbial Disease*, 2nd ed.; Wiley-Blackwell: Chichester, UK, 2011.

34. Suresh, A.K. *Co-Relating Metallic Nanoparticle Characteristics and Bacterial Toxicity*; Springer International Publishing AG: Cham, Switzerland, 2015.

35. Brown, O.R. Instrumented quantitation of live bacteria in the presence of dead cells. *Appl. Microbiol.* **1966**, *14*, 1054–1055. [PubMed]

36. Mosselhy, D.A.; Abd El-Aziz, M.; Hanna, M.; Ahmed, M.A.; Husien, M.M.; Feng, Q. Comparative Synthesis and Antimicrobial Action of Silver nanoparticles and Silver Nitrate. *J. Nanopart. Res.* **2015**, *17*, 1–10. [CrossRef]

37. Bao, H.; Yu, X.; Xu, C.; Li, X.; Li, Z.; Wei, D.; Liu, Y. New Toxicity Mechanism of Silver Nanoparticles: Promoting Apoptosis and Inhibiting Proliferation. *PLoS ONE* **2015**, *10*, e0122535. [CrossRef] [PubMed]

38. Mirzajani, F.; Ghassempour, A.; Aliahmadi, A.; Esmaeili, M.A. Antibacterial effect of silver nanoparticles on *Staphylococcus aureus*. *Res. Microbiol.* **2011**, *162*, 542–549. [CrossRef] [PubMed]

materials

MDPI

Article

Non-Equilibrium Plasma Processing for the Preparation of Antibacterial Surfaces

Eloisa Sardella [1,*, Fabio Palumbo [1], Giuseppe Camporeale [2,*] and Pietro Favia [1,2]**

[1] Istituto di Nanotecnologia, Consiglio Nazionale delle Ricerche, Via Orabona 4, 70126 Bari, Italy;
fabio.palumbo@cnr.it (F.P.); pietro.favia@uniba.it (P.F.)

[2] Dipartimento di Chimica Università degli Studi di Bari "Aldo Moro", Via Orabona 4, 70126 Bari, Italy

* Correspondence: eloisa.sardella@cnr.it (E.S.); giuseppe.camporeale1@uniba.it (G.C.);
Tel.: +39-0805-442110 (G.C.); +39-080-5442295 (E.S.)

Academic Editor: Mauro Pollini
Received: 29 April 2016; Accepted: 20 June 2016; Published: 25 June 2016

Abstract: Non-equilibrium plasmas offer several strategies for developing antibacterial surfaces that are able to repel and/or to kill bacteria. Due to the variety of devices, implants, and materials in general, as well as of bacteria and applications, plasma assisted antibacterial strategies need to be tailored to each specific surface. Nano-composite coatings containing inorganic (metals and metal oxides) or organic (drugs and biomolecules) compounds can be deposited in one step, and used as drug delivery systems. On the other hand, functional coatings can be plasma-deposited and used to bind antibacterial molecules, for synthesizing surfaces with long lasting antibacterial activity. In addition, non-fouling coatings can be produced to inhibit the adhesion of bacteria and reduce the formation of biofilm. This paper reviews plasma-based strategies aimed to reduce bacterial attachment and proliferation on biomedical materials and devices, but also onto materials used in other fields. Most of the activities described have been developed in the lab of the authors.

Keywords: antibacterial coatings; plasma processing; surface characterization

1. Introduction

1.1. Materials Related Infections and Common Antibacterial Approaches

Bacterial adhesion on biotic and abiotic surfaces is a widespread problem in many fields, and its prevention has dictated considerable research efforts during the last decades. In healthcare, the function of therapeutic and diagnostic devices can be compromised by nonspecific adsorption of proteins and adhesion of cells onto device surfaces during long-term in vivo and ex vivo exposure to physiologic fluids. A good example is the case of implanted prosthesis: upon implantation, a competition exists between integration of the device into the surrounding tissue and adhesion of bacteria at the implant surface. For successful implants, tissue integration should occur before appreciable bacterial adhesion, thereby preventing colonization at the implant. However, host defenses are often unable to prevent bacteria adhesion and colonization [1]. In this case, bacteria first adhere to the biomaterial interface, then actively bind to the extra-cellular matrix (ECM) surrounding the implant, and form a protein layer at its surface. In this process, bacteria progressively form a biofilm, i.e., a well-defined metabolic state of bacteria life cycle where microbial cells, driven by a quorum sensing mechanism, lower their growth rate and baseline metabolism and express mechanisms leading to cellular aggregation in an amorphous polysaccharide matrix, also called slime. Biofilms allow bacteria to survive in harsh environmental conditions [2]. Nosocomial infections like those related to the use of central venous catheters, urinary catheters, prosthetic heart valves, and orthopedic devices are clearly associated with biofilms adhering to biomaterial surfaces. Since bacteria in biofilms can evade host defenses and withstand antimicrobial systemic chemotherapy [3], the utility of antimicrobial coatings able to repel and/or kill bacteria appears clear.

Bacterial contamination is also a serious concern in food packaging and in several fields of industrial engineering. Antibacterial packaging materials are needed to control the microbial contamination of solid/semi-solid food by inhibiting the growth of microorganisms at its surface, which normally comes into direct contact with the packaging material. Controlling bacterial proliferation means, in the food industry, prolonging the shelf life of food without altering its organoleptic properties. It also prevents humans from being infected by food-borne diseases, mainly caused by microbiological spoilage and contamination of food with pathogen microorganisms [4,5]. It is of paramount importance that this target is achieved with materials well tolerated by the human metabolic system, with no use of allergenic or potentially toxic compounds. Since active packaging could potentially be applied to most foods, the audience interested in this technology is huge.

One of the most serious concerns in Industrial Engineering is the occurrence of biofouling and biofilms on surfaces in contact with proteins and other biologic contaminants. Biofouling is defined as the settlement and accumulation of floating organisms onto an inanimate substrate; marine biofouling is also a concern for skull and propellers of vessels [6]. In industrial settings, biofouling can often degenerate into bacterial biofilms, representing both a hygienic and an economic problem: they can contaminate the conveyor belts in the production process, reduce heat transfer and operating efficiency in heat exchange equipment, increase energy and water consumption, cause mechanical blockages and accelerate corrosion of metal surfaces [7]. Marine biofouling is also a concern for skulls and propellers of vessels.

Considering the high number of fields where surface bacterial contamination represents a severe danger, and the size of the related business, it is easy to understand why in the last decades a huge scientific and technological effort has been devoted to optimize and set up methods to prevent bacterial adhesion and proliferation on material surfaces. Accordingly, research has focused on the development of antibacterial surfaces and thin coatings that can be applied to different devices to confer resistance to bacterial colonization without affecting other properties. For instance, the visual clarity of contact lenses should not be affected by an antibacterial coating, and the flexibility of a vascular graft should not be limited by a bioactive coating on its lumen. In general, such coatings should not establish adverse effects on host bodies, cells and tissues. Moreover, given the variety of biomedical devices and implants, as well as bacteria, a single successful approach is not feasible, so antibacterial strategies need to be tailored to specific needs [8]. For contact lenses, for example, non-fouling coatings seem appropriate, as they will resist both the attachment of bacteria and biofouling [9]; on the other hand, for hip and knee prostheses, one wishes to deter bacteria for long periods of time, in order to avoid late-stage infections, while encouraging a good integration of the implant with human tissue. To achieve this target, a composite coating should be deposited at the implant surface, with the task of releasing antibacterial compounds in sub-inhibiting concentrations, and maintaining environmental drug concentrations higher than the Minimum Inhibiting Concentration (MIC) of the target organisms for a long time. This result can be achieved embedding bioactive compounds in polymeric matrices that limit their diffusion in the surrounding medium [10]. A fine level of control on the release kinetics could be acquired with hydrogel coatings, for example, from poly(ethylene glycol) di-acrylate and co-polymers; water absorption swells the polymer network and opens the way to the diffusion of the active compound [11,12]. Another common strategy for drug delivery is based on the biodegradability of the coatings: bioresorbable polymers, such as poly(lactic acid), can be degraded in newly formed human tissues into molecules that can be easily metabolized by human cells. Concurrent release of the active embedded compounds can take place while the coating is dissolved [13].

Drop casting/dip coating [1,14,15], Layer-by-Layer (LbL) deposition [16–19], sol-gel [20,21] and electrochemical deposition [11,12,22–24] are probably the most common ways for coating the surface of solid materials with a drug-delivering matrix providing a time-controlled release of bioactive ingredients (Table 1). Some processes are instead based on the covalent attachment of carboxylic, amine or thiol groups normally present in antibacterial molecules to proper surface reactive chemical groups exposed by functionalized polymeric coatings (see Table 1). Covalent immobilization has found a wide application in all technological fields requiring the permanence of the active compound on the topmost surface of the coatings, such as permanent bone prosthesis, industrial parts, nosocomial tools, etc. On the other hand, since the active ingredients are not released, these bio-conjugated surfaces cannot be effective through wide spatial ranges [25–27].

Table 1. Examples of bioactive surfaces with antibacterial properties.

Drug Delivery Systems			
Technology	**Antibacterial Agent**	**Bacterial Target**	**Ref.**
Drop Casting/Dip Coating	Chlorexidin	-	[14]
	Immunoglobulin G	*Escherichia coli*	[15]
Layer-by-Layer	Defensin	*Micrococcus luteus* *E. coli*	[16]
	PEGylated polylysine	*E. coli*	[18]
	Triclosan	*Staphylococcus aureus*	[19]
Sol-gel	Ag^+/Zn^{2+}	*Staphylococcus mutans*	[28]
	CuO	*S. aureus* *E. coli*	[29]
Electrochemical Deposition	Cu^{2+}	*S. aureus* *E. coli*	[11]
	Ag^+	*S. aureus* *Pseudomonas. Aeruginosa* *E. coli*	[12]
	Collagen	-	[22]
	Penicillin/Streptomicin	-	[23]
Covalent Immobilization			
Surface Reactive groups	**Antibacterial Ingredient**	**Bacterial target**	**Ref.**
-COOH/-F	Melimine	*P. aeruginosa* *S. aureus*	[25]
-NH$_2$	Algnic acid	-	[26]
-COCl	Vancomycin	*S. aureus*	[30]

1.2. Plasma Technology

Plasma, the "fourth state of matter", the most common state in nature, is ionized gas with equal density of positive and negative particles [31]. Thermonuclear plasmas apart, less energetic thermal (10^3–10^4 K) and cold (room temperature, RT) plasmas, populated by electrons, ions, atoms and molecules in different states, are relevant for technological applications. From their start in Microelectronics (Integrated Circuits, IC) and Photovoltaics (Solar Cells) in the 1970s, cold plasmas permeate several other industrial fields today, including the biomedical ones. Ref. [32] offers a brief Essay on the temporal evolution of plasmas in Science and Technology.

Cold (non-equilibrium) plasma discharges can be ignited at low (1–100 Pa, LP) or atmospheric pressure (AP) by applying a proper electrical field (generally in kHz, MHz or MW) to a gas/vapor mixture in properly configured plasma reactors/source. In a typical LP system, the "parallel plate" plasma reactor, the discharges is ignited between two parallel electrodes a few cm apart, with substrates positioned on one of the two. Many different configurations exist (electrodes, coils, movable substrates, etc.) to optimize plasma processes on particular substrates (flat, porous, webs, inside of narrow tubes, particles, etc.). AP systems include Dielectric Barrier Discharges (DBD) and AP plasma jets (APPJ). In DBDs, a dielectric layer (e.g., alumina) onto one or both electrodes, allow homogeneous (glow) rather than non-homogeneous (filamentary) discharges, and reduce the current (heating) through the system. An inter-electrode gap of a few mm is typical of DBD systems. In APPJs, the plasma is generated in a small caliber dielectric tube (e.g., with a metal needle electrode connected to the HV generator and a ground ring electrode wrapped around the tube), and then ejected on the substrate by the flow of the feed. Plasma reactors and sources can be precisely designed and optimized for LP and AP processes for processing small lab scale samples, or large scale high throughput industrial products [33].

By properly varying plasma parameters such as LP/AP regime, nature of the feed, reactor features and materials, frequency power and modulation of the electric field, and others, it is possible to properly tune density and distribution of the active species (atoms, radicals, ions, and electrons) generated by the fragmentation of the gas feed that interact with the exposed substrates. In this way, surface composition and properties of material substrates can be tailored in many possible ways with heterogeneous gas/solid reactions. Advantages of plasmas for materials processing include: the ability to modify thin surface layers of conventional materials with no alteration of the bulk, limited use of chemicals, dry technology with no use of solvents, easy integration in industrial processes, and intrinsic sterility. Three kinds of surface modification plasma processes can be defined, in general, as follows.

Dry etching: When defined species are generated in the plasma in the presence of certain materials (e.g., oxygen atoms vs. polymers, etc.), ablation reactions can occur, which form volatile species from the substrate (e.g., H_2O and CO_2 from polymers). By coupling cold (mostly LP) plasmas with micro/nano lithographic techniques, etching processes can be performed very precisely (e.g., on Si and other semiconductors), with lateral resolution of tens of nm and very high aspect ratio, to satisfy the needs of extreme miniaturization and newer materials for IC fabrication [34].

Plasma Enhanced Chemical Vapor Deposition (PE-CVD): Radical species are generated in LP/AP PE-CVD plasmas fed with proper chemicals to deposit coatings $1-10^3$ nm thick. Depending on the feed and on the experimental conditions, substrates can be decorated with coatings of many possible chemical compositions, including silica-like, silicone-like, fluorocarbon, diamond-like and others, for a wide range of properties (hardness, hydrophilic/phobic character, cytocompatibility, resistance to corrosion, etc.) that can be imparted at the surface of substrates. Since the chemical compound of the feed usually generates many possible fragments, building blocks of the coating, also volatile compounds without the functional groups (e.g., double bonds) needed for conventional polymerizations (e.g., methane), can originate PE-CVD coatings, whose stoichiometry can be varied in continuo with the plasma parameters. When organic compounds are used, the terms monomer for the feed and Plasma Polymerization for the technology are used.

Plasma Treatments: In these processes, materials are exposed to cold plasmas fed with reactive (e.g., O_2, N_2, H_2, NH_3, H_2O vapor, N_2/H_2, etc.) or inert (Ar, He, etc.) non-polymerizable gases. Surface modifications imparted with Plasma Treatments are extremely shallow, a few nm, and include removal of surface contaminants, surface oxidation, and surface grafting of polar groups onto polymers to impart hydrophilic character, printability, affinity with other materials, anchor groups for (bio)molecules, cytocompatibility and other properties.

Nowadays, Plasma Sciences and Technology impacts three different areas of Medicine and Biology, namely:

- Plasma Medicine, i.e., the therapeutic use of cold AP air plasmas on living tissues for non-invasive surgery, wound sterilization and healing, blood clotting, teeth bleaching, cancer treatments, and other applications [35].
- Sterilization and decontamination of materials and devices [36].
- Surface modification of biomaterials, Tissue Engineering scaffolds, biosensors and medical devices [37] aimed to optimize the response of biological entities (proteins, bacteria, cells, fluids, and tissues) in contact with the modified material and drive, and, consequently, their behavior in vitro and in vivo.

This paper describes LP/AP non-equilibrium plasma-processes developed at the lab of the authors, aimed to develop antibacterial surfaces, namely: non-fouling Poly(Ethylene Oxide) (PEO)-like coatings; surface-immobilized antibacterial molecules; nano/bio-composite coatings releasing antibacterial biomolecules; and nano-composite coatings releasing antibacterial metal ions.

2. Non-Fouling Plasma Deposited Coatings

All surfaces allow, in principle, for the adhesion of several different species. Bacterial colonization of materials, in particular, is a surface-mediated process, that poses challenges to biomaterial scientists. A promising alternative to a conventional systemic therapy against infections is represented by engineered surfaces with tailored properties aimed at locally inhibiting bacterial adhesion and proliferation on medical devices. Fouling is a main concern in the case of surfaces exposed to aqueous environments where microorganisms can bind to a surface and form a conditioning layer, which then provides an easily accessible platform for other undesired species to attach and proliferate [38]. Hydrophobic/superhydrophobic surfaces have been proposed as anti bacterial surfaces, and they can be easily fabricated by means of PE-CVD [39]. Since the first conditioning layer of proteins and biomolecules would adhere through their hydrophobic moieties with weaker (e.g., Van der Waals) forces with respect to H-bonds and other forces acting through their hydrophilic (e.g., -COOH, -NH$_2$, -OH, etc.) moieties, weaker adhesion of biofilms and bacteria and cells would result at these surfaces [40].

Hydrophilic polyethylene glycol (PEG) polymers, also referred to as polyethylene oxide (PEO), have been shown to resist to protein and cell (including bacteria) attachment [41]. Such surfaces are defined as "non-fouling" or "anti-fouling" [42–44]. This property is believed to correlate strongly with the hydration layer at the PEO surface [45], attributed to the presence of hydrophilic ether (CH$_2$-CH$_2$-O)$_n$. functionalities. These groups create a water-solvated structure which forms a liquid-like surface with highly mobile disordered molecular chains [46–48]. For protein adsorption to occur, there must be a reduction in the dehydration entropic energy associated with the removal of surface bound water [45]. Due to this effect, the tightly bound water molecules entrapped in the PEO surface through hydrogen bonds form a physical and energetic barrier that cannot be displaced by proteins and cells. The unique non-fouling properties of PEO-modified surfaces have encouraged the development of several approaches (i.e., covalent immobilization, physical adsorption, self-assembled monolayers and plasma deposition) to PEO-like thin films [44,49–58].

Kingshott et al. [59] showed that physisorbed PEO polymers do not provide long-lasting reduction in bacterial adhesion, whereas PEO chains covalently attached to a substrate are effective. About these findings Vasilev et al. [8] hypothesize that bacteria act as "mega-surfactants" with high interfacial affinity for the material surface, displacing physisorbed polymer chains from the material surface, whereas covalently surface-grafted polymer chains resist such displacement.

Compared to other methods, LP PE-CVD processes from monomers with CH$_2$CH$_2$O moieties in their structure have been widely applied as a versatile tool to impart non-fouling properties with PEO-like coatings on a large variety of substrates [44,49–51,53,55–58,60]. Three important features have to be achieved for these coatings: good adhesion to substrates, stability in water media, and high retention of the CH$_2$CH$_2$O functionalities of the monomer. This latter parameter is known as *PEO character*, and can be evaluated by measuring the relative importance of the ether carbon component C1 at ~286.5 eV of Binding Energy in the C1s X-ray Photoelectron Spectroscopy (XPS) spectrum of the coating [49,50]. High PEO character is desirable in non-fouling surfaces for proteins and cells; coatings with low PEO character, instead, promote protein and cell/bacteria adhesion and proliferation [53]. Pre-treatments in LP plasmas fed with Ar can be useful to improve the adhesion of the PE-CVD coatings; the positive-ion bombardment associated to such pre-treatments cleans the substrates from contaminations and generates nano-roughening at their surface.

We have set-up a robust LP PE-CVD process to deposit PEO-like coatings on metals, ceramics and substrates such as polystyrene (PS), polyethylene terephthalate (PET) and glass, and studied how to deposit films with high PEO character [49,50,53]. PEO-like coatings were deposited from a feed of di-ethylene glycol dimethyl ether (DEGDME) vapors and Ar at constant flow rate ratio (0.40 sccm DEGDME, 5 sccm Ar) and pressure (400 mTorr). A stainless steel plasma reactor equipped with two "parallel plate" asymmetric stainless steel electrodes was used; the small (Ø 8 cm) one was connected to a radiofrequency (RF, 13.56 MHz) generator through a matching network while the large (Ø 18 cm) electrode was connected to ground. Varying the RF power input from 5 to 15 W allowed the deposition of coatings with different chemical compositions. A high degree of monomer fragmentation was achieved in the plasma at 15 W, resulting in coatings with a PEO character of about 40% and advancing/receding water contact angles (WCA) of $71° \pm 5°$ and $50° \pm 5°$, respectively. At 5 W, coatings with a PEO character as high as 80% and more hydrophilic behavior (WCA_{adv} $56° \pm 5°$, WCA_{rec} $37° \pm 5°$) were obtained [53]. To confirm that the PEO character is correlated to the fouling/non-fouling behavior of the coatings, PE-CVD coated quartz crystals were exposed to a fibronectin (a cell adhesion protein) solution into a Quartz Crystal Microbalance (QCM-D) at 37 °C. The coatings deposited at 5 W, with high PEO-character, proved to be very stable in water solution, well adherent to substrates, with no leach of compounds and no protein adsorption from the solution. The adhesion of fibronectin was observed, instead, on coatings with lower PEO-character. The non-fouling activity of PEO-like coatings deposited at low power was confirmed by their low affinity with human telomerase cancer cells (hTERT-BJ1), keratinocytes and murine fibroblasts; cells could neither spread nor even adhere on PEO-like surfaces characterized by high PEO character [50,53].

Standing the proved non-fouling activity of PEO-like coatings plasma-deposited in mild LP plasma conditions, only few applications have been reported in literature about their use to prevent surface bacterial colonization [61–63]. The principles ruling the non-fouling properties of hydrophilic surfaces towards proteins and eukaryotic cells should also be valid for bacteria; PEO-like coatings LP plasma deposited from DEGDME should thus be able to inhibit protein and cell adhesion to solid substrates, as well as bacterial fouling. LP PE-CVD has proven to be a robust, versatile tool to produce this kind of surfaces and to adapt their characteristics to the application needs. Sardella et al. have plasma-deposited micro-patterned surfaces alternating PEO-like domains with cell adhesive ones derived from PE-CVD of acrylic acid. These coatings can drive cells to adhere only to specific regions of the substrates, leaving the non-fouling ones clean [50,53]. By combining PE-CVD with colloidal lithography, non-fouling PEO-like coatings could be generated on surfaces characterized by nanometric reliefs, that quenched the cell-adhesive properties of the rough PET substrate underneath [49].

AP PE-CVD processes have also received great attention for the deposition of PEO-like coatings [64]. Recently, the possibility to feed AP plasma processes with aerosols of liquids, of solutions or suspensions (e.g., of biomolecules or of nanoparticles, NPs) has been explored, with very interesting results. This approach allows the use of precursors with high boiling points that could not be used in LP processes or with common bubbling systems. We have studied how to deposit PEO-like films by means of aerosol-assisted AP plasma processes. A He aerosol of tetraethylene glycol dimethyl ether (TEGDME) was used as feed. Two different He flow rates were used: a constant He flow to generate the TEGDME/He aerosol with a constant flow of monomer, and a variable flow of He (carrier) to transport the aerosol. The effects of several experimental conditions on the chemical composition of the coatings were investigated, ranging from PEO-like non-fouling to cell-adhesive. Experiments were carried out in a parallel plate electrode DBD reactor. Each electrode (50×50 mm^2; 4 mm gap) was covered with a thin alumina plate. The effects of applied voltage, excitation frequency (both influencing the power of plasma) and monomer dilution could be investigated on the PEO character and on the deposition rate of the coatings. Indeed, the AP regime allowed higher deposition rates compared to corresponding LP processes. The deposition rate could be improved by increasing the applied voltage (but the PEO-character of the coating became lower); increasing the frequency led both to a lower deposition rate and to a lower PEO-character. Increasing frequency and voltage, in fact, increases the plasma power and, as in LP processes, leads to higher monomer fragmentation and lower PEO-character. A significant reduction of intact ether moieties was also observed by increasing the flow of He carrier, which lowers the concentration of the aerosol and increases the monomer fragmentation [65,66].

3. Bio-Conjugated Plasma Modified Surfaces

Covalent grafting of antibacterial compounds onto biomaterial surfaces has been the subject of considerable research in the quest for antibacterial surfaces with long-lasting efficacy. Certain PE-CVD coatings and plasma treated surfaces are well suited as adhesive interlayers for the covalent surface immobilization of antimicrobial molecules for several reasons: ease of deposition; good adherence on most materials; presence of reactive chemical groups for covalent bonding not available on the material underneath. These processes, traditionally carried out in LP plasmas, usually include two main steps, as shown in Figure 1: an initial phase aimed at enriching the topmost surface of the substrate with functional groups (grafting oxygen- or nitrogen-containing reactive groups or depositing films functionalized with the same groups); adsorption or binding (e.g., by covalent bonds) of the bioactive molecule to the surface. Plasma-synthesized surfaces rich in amine, carboxylic, epoxy and aldehyde groups have been used by many scientists for their compatibility with well-established chemical reactions for immobilizing bioactive molecules such peptides, proteins, quaternary ammonium compounds, antibiotics, etc. [8,64,67–71].

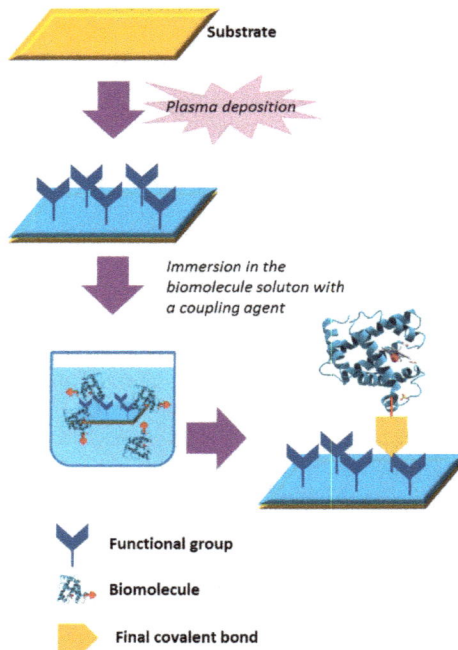

Figure 1. General scheme of a covalent immobilization on a functionalized plasma deposited coating.

Carboxyl-functionalized coatings deposited in LP plasmas have been exploited in our lab for the covalent surface attachment of bovine lactoferrin (BLF) and lactoferricin B (LfcinB), one of the sub-unities resulting from BLF digestion with pepsin. BLF and LfcinB are classified as antimicrobial peptides (AMPs), i.e., small cationic polypeptides (<10 kDa; 3–50 amino acid residues) produced by all organisms, from plants to insects to humans as a major part of their non-specific defenses against infections. They have good activity against most bacteria, and excellent activity

(MIC of 1–4 µg/mL) against multidrug resistant *Pseudomonas aeruginosa* or methicillin-resistant *Staphylococcus aureus* [72]. The antimicrobial activity of BLF and LfcinB, immobilized on the -COOH groups of a PE-CVD ethylene/acrylic acid (pdEthAA) film, was evaluated to control *Pseudomonas* strains responsible for the spoilage of mozzarella cheese [68]. This procedure was already successfully tested for the immobilization of other biomolecules or bio-structures, such as vesicles, heparin and RDG cell-adhesive peptides [73–77]. Plasma deposition of pdEthAA films was carried out in a RF plasma reactor equipped with two parallel internal and symmetric stainless steel electrodes. The upper electrode is ground-shielded and connected to a RF (13.56 MHz) generator through a matching network, whereas the lower, sample holding electrode is connected to the ground. Using mild enough experimental conditions (i.e., 30 W) and a constant 3:1 flow ratio of ethylene and acrylic acid vapors, COOH-functionalized coatings highly stable in water could be deposited in polypropylene microtubes for bacterial culture. After PE-CVD, carboxylic groups were activated with 1-Ethyl-3-[3-dimethylaminopropyl] carbodiimide hydrochloride (EDC) and incubated with BLF or LfcinB solutions to allow the covalent immobilization of both peptides in their active form, as confirmed by the reduction of the growth rate of milk spoilage bacteria *P. gessardii* and *P. fragi* after 24 h of cultivation in the modified microtubes [68].

In another approach, glycidyl methacrylate (GMA) was plasma-deposited in an AP-DBD to demonstrate that epoxy–containing interlayers can be effectively exploited for immobilizing Dispersin B (DspB), a 42 KDa protein with anti-biofouling activity [69]. The DBD source, shown in Figure 2, consisted of two high-voltage electrodes (1 mm gap) covered with alumina and a moving ground electrode. Ar was used as carrier gas, passing through a bubbler containing GMA. The plasma was ignited at 10 KHz in continuous and pulsed mode. When the discharge is pulsed, the off-time deposition can proceed through conventional polymerization reactions, thus a higher retention of functional groups in the coating can be achieved. In pulsed mode, the plasma-on time was fixed at 10 ms, while the off-time was in the range 10–80 ms. In the FTIR spectra reported in Figure 3 (top) it can be observed that coatings deposited in pulsed mode display less broad bands in the absorption regions of epoxy groups, indicating a higher retention of such groups from the monomer. Pulsing the discharge also affects the surface morphology, as highlighted by the SEM images in Figure 4 (bottom). It has been found that pulsed conditions, at the lowest duty cycle investigated (10 ms on, 80 ms off), lead to the fast deposition of adherent and smooth layers with the highest epoxy surface density, according to FTIR and XPS (18% of the total atomic carbon) data. PE-CVD GMA coatings were used to bind DspB to stainless-steel at ambient temperature; the immobilization was controlled by means of FTIR spectroscopy and then tested against biofilm forming *S. epidermidis* ATCC35984.

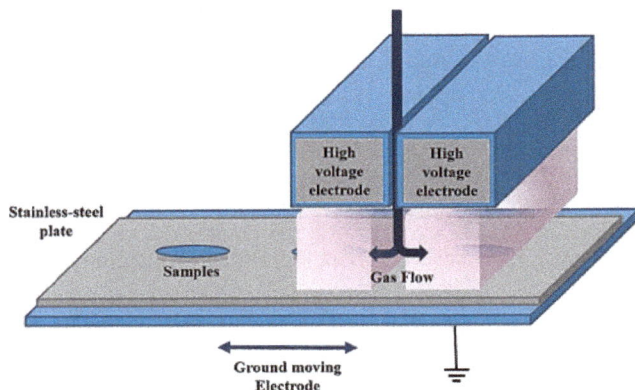

Figure 2. Schematic illustration of the AP-DBD source used for the PE-CVD from GMA, obtained from [69], with permission from © 2015 John Wiley and Sons.

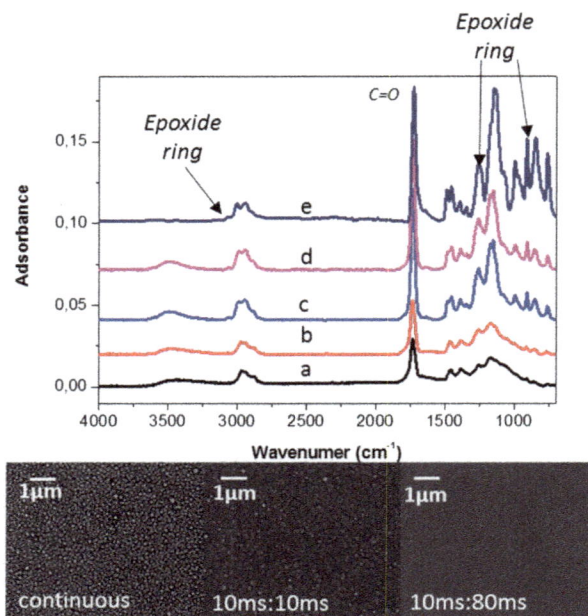

Figure 3. (**Top**) FT-IR spectra of PE-CVD GMA films deposited at 50 W at different on/off time ratio: (**a**) continuous; (**b**) 10 ms:10 ms; (**c**) 10 ms:40 ms; (**d**) 10 ms:80 ms; and (**e**) conventional GMA polymer. (**Bottom**) SEM images of PE-CVD GMA films deposited at 50 W at different on/off time ratio. Adapted from [69], with permission from © 2015 John Wiley and Sons.

The antibiofilm activity onto coated and uncoated stainless-steel substrates is reported in Table 2 as percent of reduction of the adherent bacterial population. With respect to steel substrates uncoated or coated just with PE-CVD GMA, it can be seen that Dispersine-immobilized samples led to about 80% reduction of bacteria, indicating the efficacy of the immobilization of Dispersine in its active form on the coating. Similar GMA coated surfaces were also used to bind Laccase, a hydrolase enzyme, whose activity was confirmed in the degradation of a sulfamidic antibiotic [69].

Table 2. Anti-biofouling activity against *S. epidermidis* of modified stainless-steel surfaces compared to uncoated ones. Adapted from ref. [69], with permission from © 2015 John Wiley and Sons.

Samples	Reduction of Adherent Population [%]
Stainless steel	0
ppGMA layers	0
Dsp B immobilized at pH 8.5	84 ± 11

4. Plasma Deposited Composite Coatings Embedding Organic Antibacterial Agents

When the prevention of bacterial colonization is required for short time (days–weeks), or for an entire body district and not only at the tissue/device interphase, bioactive coatings are demanded to release antibacterial drug in the surrounding medium in time- or stimuli-controlled way. Organic antibacterial compounds can be embedded in a composite coating by direct plasma deposition. The antibacterial agent can be a biomolecule with known antimicrobial activity such as lysozyme or nisin, or a synthetic drug like vancomicyn, or a commonly used quaternary ammonium salt. Essential natural oils can also be utilized [78]. The matrix can be properly chosen in order to match specific surface characteristics of the final product, e.g., biocompatibility, swelling, water resistance and so on. The drug release rate should also be controlled.

A recent approach to one-step PE-CVD of such nano/bio-composite coatings has been developed, for which a drug solution/suspension is injected in form of aerosol into a DBD through an atomizer, as shown in Figure 4 (bottom), while a polymerizable precursor is also fed. This approach is versatile, since the nature of the matrix and of the coupled biomolecule and their relative amount can be properly tuned to meet specific requirements. According to this approach, the precursor is fragmented in the discharge and plasma polymerized to form the matrix, while the active agent is simultaneously embedded within the growing film [79–81]. Damage to the biomolecule is limited by the milder fragmentation conditions typical of AP with respect to LP discharges, and by a thin protective solvent shell around the biomolecule, which can be effectively included in the coating without loss of structure and activity [80]. This concept has been proven with the optimization of Lysozyme (Lyz) containing antibacterial nano/bio-composite coatings. Ethylene was chosen as the precursor to form a hydrocarbon matrix, while Lyz was injected as a water solution aerosol [82].

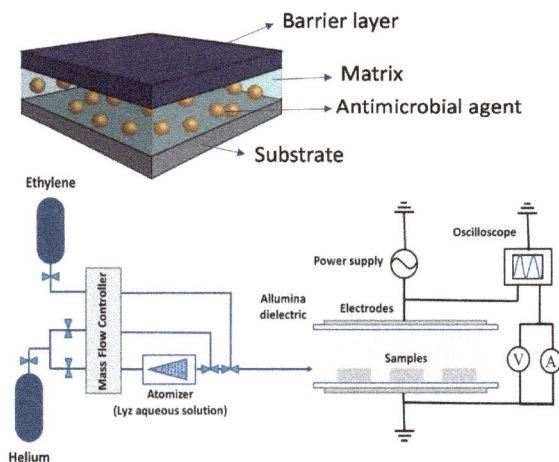

Figure 4. Sketch of the antibacterial containing nano/bio-composite coating (**top**) and schematic diagram of aerosol-assisted atmospheric-pressure DBD deposition system (**bottom**). Adapted from [82], with permission from © 2015 John Wiley and Sons.

The DBD reactor, schematized in Figure 4, consists of two parallel silver electrodes covered with 0.63 mm thick alumina plates (8×13 cm^2; 5 mm gap). The feed is pumped through the discharge by an aspirator placed on the opposite side of the flow injection. The electrode setup is confined in a sealed Plexiglas® chamber [83]. A flow rate of 2–5 slm of He passes through the atomizer (mod. 3076, TSI) to generate the Lyz water solution aerosol; 10 sccm ethylene were added to an auxiliary He inlet to keep the total flow rate at 5 slm. Crystalline polished p-doped Silicon substrates were used. A sinusoidal AC voltage was applied (6 KV$_{pp}$, 4 KHz) at a power density of 0.25 W·cm^{-2}.

In Figure 5, the absorption FTIR spectra of films deposited at various atomizer He flow rates are reported, along with the spectrum of pure lysozyme. All coatings exhibit hydrocarbon absorption bands (3015–2777 cm^{-1}, 1462 cm^{-1} and 1383 cm^{-1}). The addition of the Lyz solution results in the appearance of typical amide bands (C=O stretch at 1660 cm^{-1}, and NH bending at 1537 cm^{-1}), consistent with the presence of protein or amino acid residues in the coating. The broad band at around 3350 cm^{-1} is due to the overlap of –OH and amidic –NH$_x$. XPS analysis confirmed the presence of nitrogen in coatings prepared with the atomized Lyz solution. A deposition rate up to 75 nm/min was achieved for the coating with the highest Lyz content.

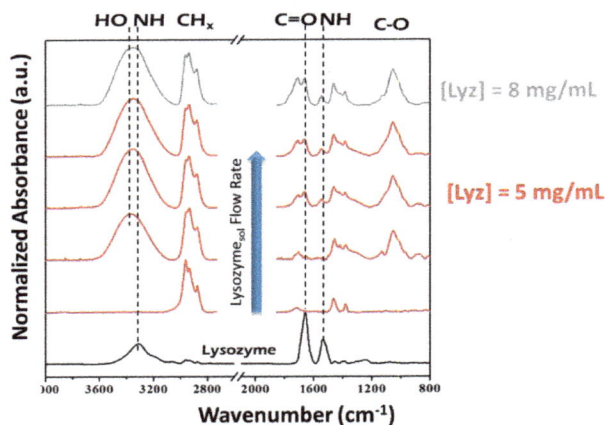

Figure 5. FT-IR spectra of: casted Lyz (black line); films deposited at atomizer flow rate of 0–5 slm with a 5 mg/mL Lyz solution (red curves); and at 5 slm atomizer flow rate with a 8 mg/mL Lyz solution (grey line). Adapted from [82], with permission from © 2015 John Wiley and Sons.

Matrix-Assisted Laser Desorption/Ionization Time-Of-Flight (MALDI-TOF, Micromass M@LDITM—LR, Waters MS Technologies, Manchester, UK; equipped with a nitrogen UV laser 337 nm) analysis was carried out to detect and identify lysozyme. The MALDI spectra of native Lyz and of Lyz-embedded coatings were similar, only showing the mono- and bi-charged ions of lysozyme at m/z 14,300 and 7150, respectively, indicating that the protein was still intact in the coating after the deposition [82].

The ability of the coatings, uniformly deposited onto 2.4 × 2.4 cm^2 silicon shards, to release lysozyme was tested in distilled water by Reverse Phase-HPLC. The chromatogram of the solution obtained from the coatings was characterized by only one signal at the same elution time of native Lyz. No protein fractions could be detected, confirming MALDI-TOF data about the absence of main alterations of Lyz. On the other side, HPLC revealed that, for the coating deposited with atomized 8 mg/mL Lyz solution, after 1 h of incubation a Lyz concentration of 28 µg/mL was found in incubation water. Further immersion led to no more release, and the FT-IR analysis of the coating revealed that all loaded Lyz was released. In order to slow down the release, samples were prepared with a 50 nm hydrocarbon barrier layer on top of the nano/bio-composite coating, deposited in the same experimental conditions, from a He 5 slm/C$_2$H$_4$ 10 sccm feed (no Lyz aerosol). The added barrier layer consistently reduced the release, as reported in Table 3. This strategy is clearly valuable for designing thin systems with programmed release rate. The nano/bio-composite coating with a 50 nm barrier layer could release Lyz for 7 days of immersion in water.

Table 3. Released Lysozyme from composite coatings deposited on 2.4 × 2.4 cm silicon shards at an atomizer flow rate of 5 slm with a 8 mg/mL Lyz solution, without and with a 50 nm hydrocarbon PE-CVD barrier layer.

Released		15 Min	1 Day	7 Days
Lysozyme [µg/mL]	Without barrier coating	20	28	-
	With barrier coating	2	15	18

The bioactivity of released lysozyme was tested with an agar diffusion test against *Micrococcus lisodeikticus*. When lysozyme diffused through the agar loaded with bacteria, agar turned from opaque to transparent, attesting for the antibacterial activity of the protein. Lysis net halos diameters are reported in Table 4. It can be observed that the solution released from the PE-CVD Lyz coating induced a halo with a diameter of about 8 mm, quite close to that originated from a 30 µg/mL standard protein solution. As expected, no halos were found for extract solutions of Lyz-free coatings. Such findings show the possibility to effectively include anti-microbial agents in PE-CVD coatings. Since the bio-activity seems to be retained, important applications can be envisioned for the dry manufacturing of drug carrier systems with programmed delivery.

Table 4. Agar diffusion activity test results for the Lyz-containing coating from a composite layer prepared with a 8 mg/mL solution aerosol, at 5 slm of He. Based on [82].

Sample	Inhibition Halo Diameter [mm]
C_2H_4/Lyz_{sol} coating	8 ± 1
C_2H_4/H_2O plasma deposited coating (control)	0
Lyz std solution (10 µg/mL)	0
Lyz std solution (30 µg/mL)	6 ± 1
Lyz std solution (300 µg/mL)	12 ± 1
Blank (negative control)	0

5. Plasma Deposited Composite Coatings Embedding Inorganic Antibacterial Agents

Certain metals are known to possess bactericidal properties, among them silver, titanium dioxide (TiO_2), zinc oxide (ZnO) and copper oxide (CuO) have enabled the development of a new generation of biocides. Depending on the metal property, the biocide behavior of metals and metals oxides can be triggered by different mechanisms including the involvement of metals as catalytic co-factors in the generation of reactive oxygen species (ROS) [84]. In this mechanism, metals are involved both in cell oxidative stress and in triggering pro-inflammatory signal cascades that promote programmed cell death [85]. Another possible mechanism of action is based on the binding of the metals to atoms (e.g., O, N, and S), of donor ligands through strong and selective interactions that inhibit vital biological roles covered by the donors [86]. Metals and metal oxides can exert their function either in the cell membrane or in the intracellular region [87,88]. Generally, the biocidal activity of such inorganic antimicrobials has been found to be associated to the ionic form [89].

The antibacterial activity of silver and silver ions in proper concentrations, accompanied by low toxicity to human cells, is well known since a long time ago [90]; in addition, bacterial resistance is not developed. Silver has been incorporated into the surface of a variety of medical devices, such as vascular, urinary and peritoneal catheters, vascular grafts, prosthetic heart valve sewing rings, surgical sutures and fracture-fixation devices [91]. During the last years, nanotechnology has produced a new route to take advantage of the antimicrobial behavior of metals by synthesizing highly active metal nanoparticles (NP) [92].

Several surface modification approaches aimed at introducing inorganic antibacterial elements as well as NPs have been described so far, and comprehensive reviews have been published [8,20,93]: a scheme of the possible strategies for plasma-assisted approaches used to synthesize coatings containing antibacterial inorganic elements is shown in Figure 6.

One of the plasma-assisted strategies, illustrated in Figure 6a–c, requires the plasma activation of the material surface and further exposure of the material to a solution of metal compounds or of NPs. The plasma process is used to improve the interaction between the metal (or NPs) and the surface, or for the further modification of the adsorbed metal (i.e., chemical reduction, and production of NPs). Leys et al. [94,95] described a three-step approach (Figure 6b) consisting of the AP PE-CVD of a silicone-like thin film using an AP plasma jet, followed by the dipping-adsorption of silver nano-particles (AgNPs), then by a final AP-PE-CVD of a silicone-like barrier layer that allowed the reduction and control of the release of Ag$^+$ ions. Such coatings show antimicrobial activity against *P. auruginosa*, *S. aureus*, and *C. albicans*. A similar strategy was performed also on nylon plasma-treated with a DBD in atmospheric air [96].

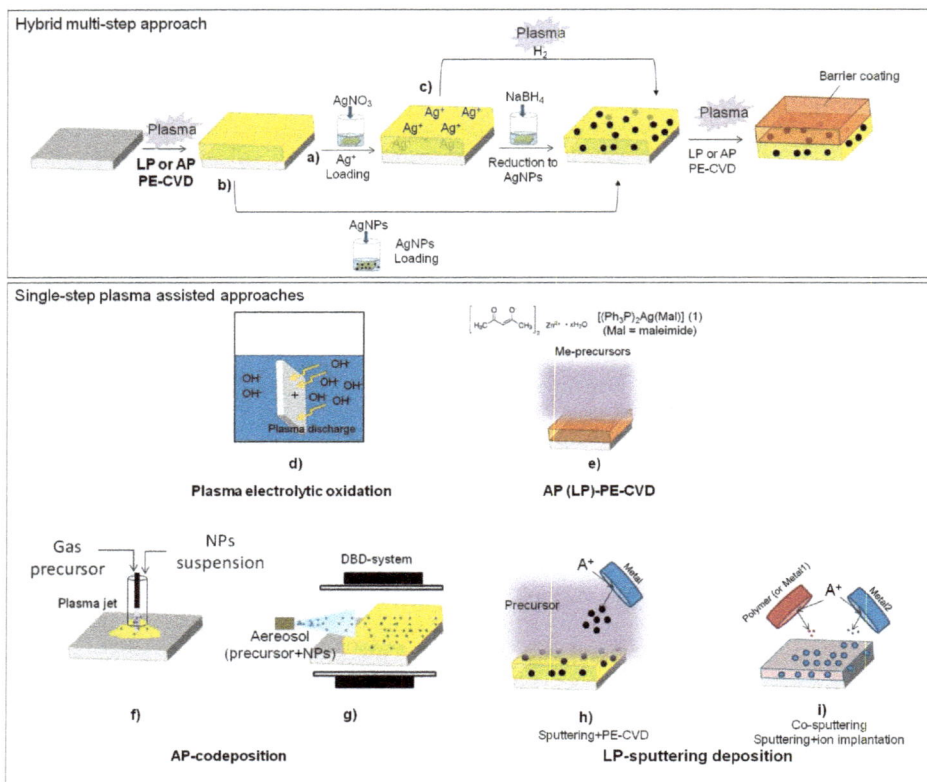

Figure 6. Scheme of plasma assisted approaches. The first route is a hybrid multi-step approach consisting of a PE-CVD coating containing carboxylic or amino groups that enhances (**a**) the adsorption of Ag$^+$ ions or (**b**) of AgNPs at the surface and/or inside the coating. The reduction of Ag$^+$ ions with NaBH$_4$ (or trisodium citrate) or (**c**) through a plasma fed with H$_2$ occurs. Finally, a PE-CVD barrier coating allows the control of silver release. The second group of processes are single step approaches totally based on plasma processes; (**d**) Electrolytic anodization is performed at high potential to ignite a discharge at material/liquid interface to change the chemical state of the surface and embedd antibacterial components (e.g., AgNO$_3$) within its surface; (**e**) PE-CVD of a metal-containing volatile precursor (e.g., zinc acetate or silver bisphosphine, or maleimide complex) can be used to deposit metal-containing nanocomposite coatings; a plasma co-deposition of NPs and of an organic precursor allow to obtain a nanocomposite coating both in case of (**f**) an AP Plasma Jet or of (**g**) a DBD fed with an aereosol of a suspension of NPs; (**h**) A process based on plasma sputtering of a metal of interest and PE-CVD from a precursor is aimed at obtaining a metal-containing nanocomposite coating; (**i**) A contemporary sputtering of a polymer target and a metal one, or a deposition of the polymer of interest by PE-CVD and a contemporary ion implantation can lead to a thin layer containing nanoclusters of the metal of interest dispersed in an organic (or inorganic) matrix. Abbreviations: Mal: maleimide; Ph: phosphine.

Hydrogel-like plasma deposited coatings (Figure 6a) have been used to fix Ag^+ ions from an $AgNO_3$ solution. The authors demonstrated control over the release of Ag^+ by depositing a thin (<20 nm) plasma-polymer overlayer [97]. Following this strategy, some research groups proposed the PE-CVD of coatings from a proper organic precursor (i.e., acrylic acid or n-heptylamine) to absorb Ag^+ ions from an $AgNO_3$ solution and to reduce them with $NaBH_4$ to form AgNPs embedded in the coating [98,99]. Such surfaces were found to be bioactive against *E. coli* and *S. aureus*.

Plasma-deposited PEO-like coatings exhibit hydrogel behavior in water [100]; for this reason, they have been investigated by the authors to absorb Ag^+ ions to be reduced to metal silver. PEO-like coatings were plasma-deposited at LP [101] at 60 W power (high fragmentation conditions), 20 sccm Ar, 0.25 sccm DEGDME, 50 mTorr for 15 min. XPS show high cross-linking of the coating, attested by the low amount of $-CH_2CH_2O-$ moieties (i.e., low PEO-character, grey peak in Figure 7), and by the presence of -COOH groups. These coatings were chosen to uptake Ag^+ ions through ionic bonds with their ionized -COOH groups. Following the procedure of ref. [102], the samples were rinsed for 15 min in solutions at pH 4 (HCl 10^{-4} M) and pH 8 (NaOH 10^{-6} M), then they were exposed for 15 h to a 1 mg/mL solution of $AgNO_3$ to absorb Ag^+ ions, and finally immersed in a water solution of trisodium citrate (0.7 mg/mL) for 24 h for reducing Ag^+ to Ag. As shown in Figure 7, the coatings conditioned at pH 8 were able to embed a higher amount of silver, likely due to the interactions of Ag^+ with the surface carboxylate ions formed at higher extent than at pH 4. The concentration of silver detected is similar to that reported by Kumar et al. [99] who showed an abatement of bacterial concentration higher than 99.7% compared to untreated PET against *E. coli* and *S. aureus*. The advantage of such approach consists in the possibility to control the loading of silver by controlling the chemical composition of the deposited material and the pH; the disadvantage is that at least two steps are necessary to obtain such coatings. To overcome this limit, one can incorporate AgNPs into a PE-CVD coating by a simultaneous deposition of a metal (metal oxide) and a plasma organic or inorganic matrix (Figure 6d–i). There are at least three approaches, uniquely based on plasma processing, to perform such depositions: (1) plasma electrolytic oxidation (Figure 6d); (2) plasma deposition from a metal containing precursor (Figure 6e); and (3) dual strategies (Figure 6f–i).

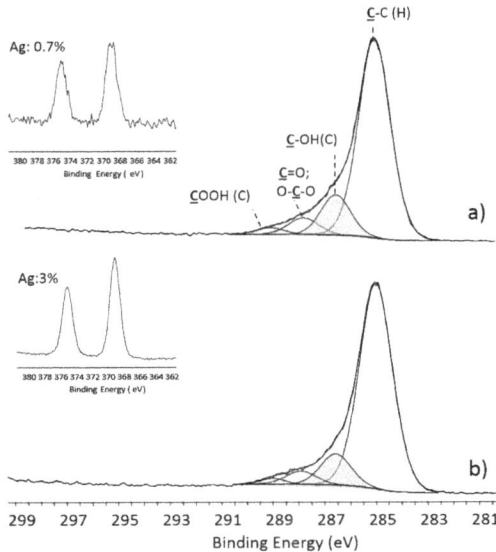

Figure 7. XPS spectra acquired on silver loaded PEO-like coatings after pre-conditioning of the coating at: (**a**) pH 4; and (**b**) pH 8.

A new surface modification approach involves the electrochemical conversion of metal surfaces (i.e., Titanium) into more desirable chemistries and topographies by using anodizing or plasma electrolytic oxidation processes (Figure 6d) [103,104]. This technique produces coatings characterized by an excellent adhesion to substrates; however, it requires the use of a metal substrate. To overcome this limit, hybrid processes where a plasma-assisted approach is coupled with another one are proposed. The most frequent examples of hybrid plasma strategies are reported in Figure 6e, PE-CVD from a precursor containing a complex of the metal of interest or organo-metal compound [105,106]; Figure 6f,g AP plasma co-deposition of an organic coating and AgNPs or AgNPs precursors [107–109]; and Figure 6h,i, simultaneous sputtering of a target coated with the antibacterial agent (i.e., Cu, Ag, TiO_2) and PE-CVD of an organic precursor or ion beam deposition [110,111].

Polymer-like coatings represent a potentially interesting approach to controlled drug release of inorganic antimicrobial compounds. Plasma polymerization of an organic film while adding an antibacterial agent by sputtering from a target is one of the most widely used and well characterized approach in this field. Silicone-like/Ag [112], silicone-like/Cu, diamond-like/Ag, ethylene-like/Ag [113] and PEO-like/Ag nanocomposite coatings have been produced in this way. The formation of Ag clusters in different matrices is promoted by the high mobility of Ag atoms in the growing coating, which leads to aggregation of Ag clusters [114], whose size is strongly dependent on the Ag content [115]; often, the silver distribution is not uniform along the thickness of the coatings [116].

To deposit Ag/PEO-like nanocomposite coatings we used a LP deposition system with parallel asymmetrical plate electrodes. The smallest one (7 cm dia) covered with a 2 mm thick Ag lamina, is connected through a matching network to a RF (13.56 MHz) power supply, the other one (25 cm dia), sample holder, is grounded. The asymmetric configuration allows a strong positive ion-bombardment on the silver electrode, at proper pressure, power and gas composition, that sputters Ag atoms from the target. At the same time a PEO-like coating develops at the ground electrode, where the ion bombardment is reduced, that incorporate Ag atoms and nanometric clusters. DEGDME vapors were flown at 0.25 sccm in all experiments, mixed with a variable flow of Ar (10–20 sccm). The input power was changed from 10 to 60 W at constant pressure (50 mTorr) to vary the amount of sputtered silver. In these conditions, the deposition of a coating at the silver electrode, that would block the sputtering of Ag atoms, is prevented. By adjusting the process parameters, the coatings could be tailored (e.g., Ag content, dimensions distribution and morphology of Ag clusters) to different requirements for specific applications, where they act as Ag reservoirs. Bacterial adhesion on such samples was evaluated both in static and under flow conditions in order to investigate the combined effect of flow and surface chemistry on the ica gene expression [117]. By a comparison between Ag/PEO-like coatings and highly cross-linked PEO-like ones, but with no silver embedded, prepared in the same conditions, only 30% of adherent bacteria on Ag/PEO-like coatings remained alive. It is known that assessment of icaA DBC operon gene expression is crucial to the understanding of the pathogenesis of *S. epidermidis* infections [118]. Higher expressions of icaA genes were observed for bacteria in contact with Ag/PEO-like coatings with respect to those on the corresponding highly cross-linked PEO-like ones with no silver.

In order to preserve the antibacterial efficiency while protecting eukaryotic cells from the toxic effect of the excessive Ag^+ release, a barrier layer was deposited on top of plasma deposited Ag/PEO-like coatings [119] to limit the release of Ag^+ ions in water media (Figure 8). Thickness and chemical composition of the barrier significantly influence the release that may occur through small cracks and pores, as well as by diffusion through the coating and the barrier together, slightly swollen in water media [98,109]. The barrier layer can control and prolong the antibacterial effect.

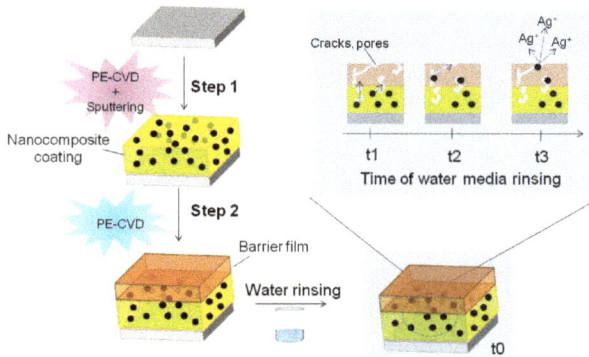

Figure 8. The scheme of silver ions generation and diffusion into a liquid medium during time starting from soon after the plasma processes. In the scheme, the first layer acts as the reservation layer while the outermost layer is used to control diffusion of silver through the coating. The generation of silver ions pass through the oxidation of silver when is in the elemental state [119].

We have tested the Ag^+ release in cell culture Dulbecco's Modified Eagle Medium (DMEM) during eukaryotic cell culture experiments, a very complex environment were the release could be altered by protein adsorption and damage of the coating induced by adhering cells. After 24, 48 and 72 h of culture, 500 µL of medium were digested in 2 mL of HNO_3 (70%, Sigma Aldrich, St. Louis, MO, USA) to oxidize the organic matter, then dissolved in 2% v/v HNO_3 solution in water. The amount of silver released was measured with ICP Atomic Spectroscopy. Ag/PEO-like coatings containing 5% of silver, as detected by XPS, were coated with 15 nm and 60 nm thick PEO-like barrier coatings with 20% PEO-character. An abatement of the released silver resulted with the 60 nm barrier, at all incubation times investigated, while no effect was observed with the 15 nm barrier. At 24 h the amount of released silver from samples coated with 60 nm overlayers was 0.5 ± 0.02 ppm·cm^{-2}, after 72 h the same coatings released 1.5 ± 0.2 ppm·cm^{-2} while the release was up to 5 ± 0.5 ppm·cm^{-2} without the barrier. When at least 1 cm^2 samples were used, independently on the barrier, the amount of released silver was higher than the MIC of Ag^+ ions (0.9–1.7 ppm) for most gram negative and positive bacteria [120].

Ag/PEO-like coatings prepared by us containing from 1% [101] to 25% of silver [101,117], show antibacterial effects against *P. aeuruginosa*, *S. epidermis*, and *E. coli*; coatings with 3% silver showed a 100% reduction of *P. aeuruginosa* adhesion, as reported in Figure 9.

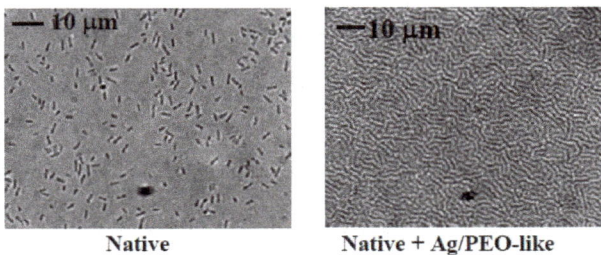

Native   Native + Ag/PEO-like

Figure 9. Micrographs of *P. aeuruginosa* adhesion onto native PVC Endotracheal Tubes as well (native) and coated by Ag/PEO-like coatings containing 3% of silver. The mosaic pattern seen on the Ag/PEO-like micrograph is due to the wrinkling of the underlying PVC substrate, adapted from [101], with permission of © 2005 John Wiley and Sons.

To check the toxicity of Ag/PEO-like coatings for eukaryotic cells, in vitro cito-toxicity tests were performed with SAOS2 cells grown on coatings loaded with 5% of silver, with and without barrier layers. Coatings without the barrier resulted somewhat toxic, and could promote the adhesion of only few round shaped cells, due to the high release of Ag^+ but, maybe, also to the direct contact of cells with Ag clusters protruding from the surface. On samples with a 60 nm thick barrier coating (20% PEO character, the same of the matrix of the Ag/PEO material), instead, more cells could adhere and spread, as shown in Figure 10, indicating a healthier environment for the cells due to the lowered release of Ag^+.

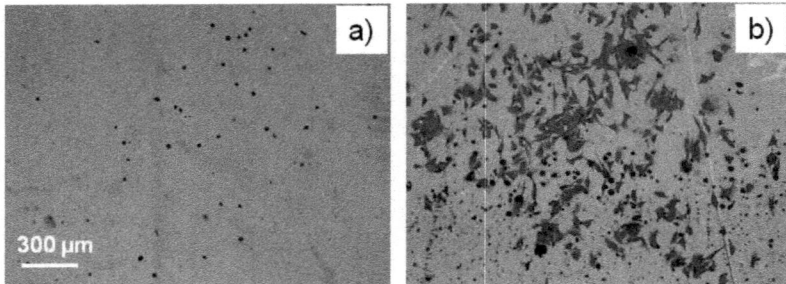

Figure 10. SAOS 2 Coomassie Blue stained cells on Ag/PEO-like (5% Ag) surfaces uncoated (**a**) and coated (**b**) with a 60 nm thick barrier coating (PEO character 20%).

6. Conclusions and Perspectives

An overview on plasma-based technologies optimized for the deposition of antimicrobial coatings has been given in this paper, where most of the surfaces described have been developed in our laboratory. In particular, we believe to have shown the power and the flexibility of LP and AP plasma-based techniques in engineering antibacterial coatings for the prevention of infections in biomedical materials, with approaches that could probably also be exploited in other fields, such as active food packaging, marine fouling and industrial food processing, where bacterial attachment and contamination, and biofilm formation are of concern.

Plasma Technology, indeed, is nowadays expressing several newer approaches to surface modification, decontamination/sterilization methods and medical therapies in several fields where it was not considered before, where of bacteria proliferation always has to be controlled, e.g., in Medicine for wound healing and cancer treatments [121], and in Agriculture for seed germination and food decontamination [122].

Among several approaches presented, in particular, the aerosol-assisted AP plasma synthesis of nano/bio-composite coatings with antibacterial molecules embedded in, and releasable from, plasma-deposited nanometric coatings, possibly implemented by plasma-deposited diffusive layers is considered, in perspective, among the most promising approaches to antibacterial surfaces and to drug-release systems in general. We believe that this synthetic method will be strongly implemented, in the near future, with the increased use of natural molecules, e.g., available from extracts and oils from plants, flowers and fruits, for an ecologic approach to antibacterial material and surfaces.

Acknowledgments: P. Choquet and M. Moreno (Luxembourg Institute of Science and Technology, LUX), D. Calvano, A. Mangone and L. Giannossa (Univ. Bari), G. Da Ponte and S. Paulussen (VITO, BEL), Y. Yang and J. Wu (Nat. Chiao Tung Univ., TAIWAN), L. Quintieri, L. Caputo and F. Baruzzi (ISPA-CNR) are kindly acknowledged for the strict collaboration on this topic. Savino Cosmai (NANOTEC-CNR) and Danilo Benedetti (Univ. Bari) are acknowledged for their valuable technical assistance. The projects LIPP (Rete di Laboratorio 51, Regione Puglia) SISTEMA (PON MIUR), RINOVATIS (PON MIUR), and MAIND (PON MIUR) are acknowledged for funds.

Conflicts of Interest: The authors declare no conflict of interest.

References

1. Hetrick, E.M.; Schoenfisch, M.H. Reducing implant-related infections: Active release strategies. *Chem. Soc. Rev.* **2006**, *35*, 780–789. [CrossRef] [PubMed]
2. Campoccia, D.; Montanaro, L.; Speziale, P.; Arciola, C.R. Antibiotic-loaded biomaterials and the risks for the spread of antibiotic resistance following their prophylactic and therapeutic clinical use. *Biomaterials* **2010**, *31*, 6363–6377. [CrossRef] [PubMed]
3. Stewart, P.S.; Costerton, J.W. Antibiotic resistance of bacteria in biofilms. *Lancet* **2001**, *358*, 135–138. [CrossRef]
4. Cutter, C.N.; Willett, J.L.; Siragusa, G.R. Improved antimicrobial activity of nisin-incorporated polymer films by formulation change and addition of food grade chelator. *Lett. Appl. Microbiol.* **2001**, *33*, 325–328. [CrossRef] [PubMed]
5. Gao, C.; Yan, T.; Du, J.; He, F.; Luo, H.; Wan, Y. Introduction of broad spectrum antibacterial properties to bacterial cellulose nanofibers via immobilising ε-polylysine nanocoatings. *Food Hydrocoll.* **2014**, *36*, 204–211. [CrossRef]
6. Chambers, L.D.; Stokes, K.R.; Walsh, F.C.; Wood, R.J.K. Modern approaches to marine antifouling coatings. *Surf. Coat. Technol.* **2006**, *201*, 3642–3652. [CrossRef]
7. Shi, X.; Zhu, X. Biofilm formation and food safety in food industries. *Trends Food Sci. Technol.* **2009**, *20*, 407–413. [CrossRef]
8. Vasilev, K.; Griesser, S.S.; Griesser, H.J. Antibacterial surfaces and coatings produced by plasma techniques. *Plasma Process. Polym.* **2011**, *8*, 1010–1023. [CrossRef]
9. Thissen, H.; Gengenbach, T.; du Toit, R.; Sweeney, D.F.; Kingshott, P.; Griesser, H.J.; Meagher, L. Clinical observations of biofouling on PEO coated silicone hydrogel contact lenses. *Biomaterials* **2010**, *31*, 5510–5519. [CrossRef] [PubMed]
10. Procaccini, R.A.; Studdert, C.A.; Pellice, S.A. Silver doped silica-methyl hybrid coatings. Structural evolution and antibacterial properties. *Surf. Coat. Technol.* **2014**, *244*, 92–97. [CrossRef]
11. Cometa, S.; Iatta, R.; Ricci, M.A.; Ferretti, C.; De Giglio, E. Analytical characterization and antimicrobial properties of novel copper nanoparticle-loaded electrosynthesized hydrogel coatings. *J. Bioact. Compat. Polym.* **2013**, *28*, 508–522. [CrossRef]
12. De Giglio, E.; Cafagna, D.; Cometa, S.; Allegretta, A.; Pedico, A.; Giannossa, L.C.; Sabbatini, L.; Mattioli-Belmonte, M.; Iatta, R. An innovative, easily fabricated, silver nanoparticle-based titanium implant coating: Development and analytical characterization. *Anal. Bioanal. Chem.* **2013**, *405*, 805–816. [CrossRef] [PubMed]
13. Lucke, M.; Schmidmaier, G.; Sadoni, S.; Wildemann, B.; Schiller, R.; Haas, N.P.; Raschke, M. Gentamicin coating of metallic implants reduces implant-related osteomyelitis in rats. *Bone* **2003**, *32*, 521–531. [CrossRef]
14. Kim, W.H.; Lee, S.B.; Oh, K.T.; Moon, S.K.; Kim, K.M.; Kim, K.N. The release behavior of CHX from polymer-coated titanium surfaces. *Surf. Interface Anal.* **2008**, *40*, 202–204. [CrossRef]
15. Rojas, I.A.; Slunt, J.B.; Grainger, D.W. Polyurethane coatings release bioactive antibodies to reduce bacterial adhesion. *J. Control. Release* **2000**, *63*, 175–189. [CrossRef]
16. Etienne, O.; Picart, C.; Taddei, C.; Haikel, Y.; Dimarcq, J.L.; Schaaf, P.; Voegel, J.C.; Ogier, J.A.; Egles, C. Multilayer polyelectrolyte films functionalized by insertion of defensin: A new approach to protection of implants from bacterial colonization. *Antimicrob. Agents Chemother.* **2004**, *48*, 3662–3669. [CrossRef] [PubMed]
17. Amorosi, C.; Michel, M.; Avérous, L.; Toniazzo, V.; Ruch, D.; Ball, V. Plasma polymer films as an alternative to (PSS-PAH) n or (PSS-PDADMAC) n films to retain active enzymes in exponentially growing polyelectrolyte multilayers. *Colloids Surfaces B Biointerfaces* **2012**, *97*, 124–131. [CrossRef] [PubMed]
18. Boulmedais, F.; Frisch, B.; Etienne, O.; Lavalle, P.; Picart, C.; Ogier, J.; Voegel, J.C.; Schaaf, P.; Egles, C. Polyelectrolyte multilayer films with pegylated polypeptides as a new type of anti-microbial protection for biomaterials. *Biomaterials* **2004**, *25*, 2003–2011. [CrossRef] [PubMed]
19. Micelles, L.B.C.; Nguyen, P.M.; Zacharia, N.S.; Verploegen, E.; Hammond, P.T. Extended Release Antibacterial Layer-by-Layer Films Incorporating. *Mater. Sci.* **2007**, *02139*, 8932–8936.
20. Simchi, A.; Tamjid, E.; Pishbin, F.; Boccaccini, A.R. Recent progress in inorganic and composite coatings with bactericidal capability for orthopaedic applications. *Nanomed. Nanotechnol. Biol. Med.* **2011**, *7*, 22–39. [CrossRef] [PubMed]

21. Barbé, C.J.; Kong, L.; Finnie, K.S.; Calleja, S.; Hanna, J.V.; Drabarek, E.; Cassidy, D.T.; Blackford, M.G. Sol-gel matrices for controlled release: From macro to nano using emulsion polymerisation. *J. Sol-Gel Sci. Technol.* **2008**, *46*, 393–409. [CrossRef]

22. Fan, Y.; Duan, K.; Wang, R. A composite coating by electrolysis-induced collagen self-assembly and calcium phosphate mineralization. *Biomaterials* **2005**, *26*, 1623–1632. [CrossRef] [PubMed]

23. Wang, H.; Lin, C.J.; Hu, R.; Zhang, F.; Lin, L.W. A novel nano-micro structured octacalcium phosphate/protein composite coating on titanium by using an electrochemically induced deposition. *J. Biomed. Mater. Res. A* **2008**, *87*, 698–705. [CrossRef] [PubMed]

24. Wang, J.; Layrolle, P.; Stigter, M.; De Groot, K. Biomimetic and electrolytic calcium phosphate coatings on titanium alloy: Physicochemical characteristics and cell attachment. *Biomaterials* **2004**, *25*, 583–592. [CrossRef]

25. Chen, R.; Cole, N.; Willcox, M.D.P.; Park, J.; Rasul, R.; Carter, E.; Kumar, N. Synthesis, characterization and in vitro activity of a surface-attached antimicrobial cationic peptide. *Biofouling* **2009**, *25*, 517–524. [CrossRef] [PubMed]

26. Yoshioka, T.; Tsuru, K.; Hayakawa, S.; Osaka, A. Preparation of alginic acid layers on stainless-steel substrates for biomedical applications. *Biomaterials* **2003**, *24*, 2889–2894. [CrossRef]

27. Dettin, M.; Bagno, A.; Gambaretto, R.; Iucci, G.; Conconi, M.T.; Tuccitto, N.; Menti, A.M.; Grandi, C.; Di Bello, C.; Licciardello, A.; et al. Covalent surface modification of titanium oxide with different adhesive peptides: Surface characterization and osteoblast-like cell adhesion. *J. Biomed. Mater. Res. A* **2009**, *90*, 35–45. [CrossRef] [PubMed]

28. Chung, R.J.; Hsieh, M.F.; Huang, C.W.; Perng, L.H.; Wen, H.W.; Chin, T.S. Antimicrobial effects and human gingival biocompatibility of hydroxyapatite sol-gel coatings. *J. Biomed. Mater. Res. B Appl. Biomater.* **2006**, *76*, 169–178. [CrossRef] [PubMed]

29. Ye, J.; He, J.; Wang, C.; Yao, K.; Gou, Z. Copper-containing mesoporous bioactive glass coatings on orbital implants for improving drug delivery capacity and antibacterial activity. *Biotechnol. Lett.* **2014**, *36*, 961–968. [CrossRef] [PubMed]

30. Kurczewska, J.; Sawicka, P.; Ratajczak, M.; Gajęcka, M.; Schroeder, G. Vancomycin-modified silica: Synthesis, controlled release and biological activity of the drug. *Int. J. Pharm.* **2015**, *486*, 226–231. [CrossRef] [PubMed]

31. Langmuir, I. Oscillations inionied gases. *Phys. Rev.* **1928**, *14*, 627–637.

32. D'Agostino, R.; Favia, P.; Oehr, C.; Wertheimer, M.R. Low-temperature plasma processing of materials: Past, present, and future. *Plasma Process. Polym.* **2005**, *2*, 7–15. [CrossRef]

33. Plasma 2010 Committee; Plasma Science Committee; National Research Council. *Plasma Science: Advancing Knowledge in the National Interest*; The National Academies Press: Washington, DC, USA, 2007.

34. Donnelly, V.M.; Kornblit, A. Plasma etching: Yesterday, today, and tomorrow. *J. Vac. Sci. Technol. A* **2013**, *31*. [CrossRef]

35. Fridman, G.; Friedman, G.; Gutsol, A.; Shekhter, A.B.; Vasilets, V.N.; Fridman, A. Applied plasma medicine. *Plasma Process. Polym.* **2008**, *5*, 503–533. [CrossRef]

36. Laroussi, M. Low temperature plasma-based sterilization: Overview and state-of-the-art. *Plasma Process. Polym.* **2005**, *2*, 391–400. [CrossRef]

37. Sardella, E.; Favia, P.; Gristina, R.; Nardulli, M.; d'Agostino, R. Plasma-aided micro- and nanopatterning processes for biomedical applications. *Plasma Process. Polym.* **2006**, *3*, 456–469. [CrossRef]

38. Banerjee, I.; Pangule, R.C.; Kane, R.S. Antifouling coatings: Recent developments in the design of surfaces that prevent fouling by proteins, bacteria, and marine organisms. *Adv. Mater.* **2011**, *23*, 690–718. [CrossRef] [PubMed]

39. Milella, A.; Di Mundo, R.; Palumbo, F.; Favia, P.; Fracassi, F.; d'Agostino, R. Plasma nanostructuring of polymers: Different routes to superhydrophobicity. *Plasma Process. Polym.* **2009**, *6*, 460–466. [CrossRef]

40. Di Mundo, R.; Gristina, R.; Sardella, E.; Intranuovo, F.; Nardulli, M.; Milella, A.; Palumbo, F.; D'Agostino, R.; Favia, P. Micro-/nanoscale structuring of cell-culture substrates with fluorocarbon plasmas. *Plasma Process. Polym.* **2010**, *7*, 212–223. [CrossRef]

41. Harris, L.G.; Tosatti, S.; Wieland, M.; Textor, M.; Richards, R.G. Staphylococcus aureus adhesion to titanium oxide surfaces coated with non-functionalized and peptide-functionalized poly(L-lysine)-grafted-poly(ethylene glycol) copolymers. *Biomaterials* **2004**, *25*, 4135–4148. [CrossRef] [PubMed]

42. Feng, W.; Zhu, S.; Ishihara, K.; Brash, J.L. Protein resistant surfaces: Comparison of acrylate graft polymers bearing oligo-ethylene oxide and phosphorylcholine side chains. *Biointerphases* **2006**, *1*, 50. [CrossRef] [PubMed]

43. Choukourov, A.; Gordeev, I.; Polonskyi, O.; Artemenko, A.; Hanyková, L.; Krakovský, I.; Kylián, O.; Slavínská, D.; Biederman, H. Polyethylene (ethylene oxide)-like plasma polymers produced by plasma-assisted vacuum evaporation. *Plasma Process. Polym.* **2010**, *7*, 445–458. [CrossRef]

44. Brétagnol, F.; Lejeune, M.; Papadopoulou-Bouraoui, A.; Hasiwa, M.; Rauscher, H.; Ceccone, G.; Colpo, P.; Rossi, F. Fouling and non-fouling surfaces produced by plasma polymerization of ethylene oxide monomer. *Acta Biomater.* **2006**, *2*, 165–172. [CrossRef] [PubMed]

45. Morra, M. On the molecular basis of fouling resistance. *J. Biomater. Sci. Polym. Ed.* **2000**, *11*, 547–569. [CrossRef] [PubMed]

46. Chen, S.; Li, L.; Zhao, C.; Zheng, J. Surface hydration: Principles and applications toward low-fouling/nonfouling biomaterials. *Polymer (Guildf)* **2010**, *51*, 5283–5293. [CrossRef]

47. Griesser, H.J.; Hall, H.; Jenkins, T.A.; Griesser, S.S.; Vasilev, K. *Intelligent Surfaces in Biotechnology*; John Wiley & Sons Inc.: Hoboken, NJ, USA, 2012; pp. 183–241.

48. Favia, P.; d'Agostino, R. Plasma treatments and plasma deposition of polymers for biomedical applications. *Surf. Coat. Technol.* **1998**, *98*, 1102–1106. [CrossRef]

49. Sardella, E.; Detomaso, L.; Gristina, R.; Senesi, G.S.; Agheli, H.; Sutherland, D.S.; d'Agostino, R.; Favia, P. Nano-structured cell-adhesive and cell-repulsive plasma-deposited coatings: Chemical and topographical effects on keratinocyte adhesion. *Plasma Process. Polym.* **2008**, *5*, 540–551. [CrossRef]

50. Sardella, E.; Gristina, R.; Ceccone, G.; Gilliland, D.; Papadopoulou-Bouraoui, A.; Rossi, F.; Senesi, G.S.; Detomaso, L.; Favia, P.; d'Agostino, R. Control of cell adhesion and spreading by spatial microarranged PEO-like and pdAA domains. *Surf. Coat. Technol.* **2005**, *200*, 51–57. [CrossRef]

51. Palumbo, F.; Favia, P.; Vulpio, M.; D'Agostino, R. RF plasma deposition of PEO-like films: Diagnostics and process control. *Plasmas Polym.* **2001**, *6*, 163–174. [CrossRef]

52. Lee, J.H.; Lee, H.B.; Andrade, J.D. Blood compatibility of polyethylene oxide surfaces. *Prog. Polym. Sci.* **1995**, *20*, 1043–1079. [CrossRef]

53. Sardella, E.; Gristina, R.; Senesi, G.S.; D'Agostino, R.; Favia, P. Homogeneous and micro-patterned plasma-deposited PEO-like coatings for biomedical surfaces. *Plasma Process. Polym.* **2004**, *1*, 63–72. [CrossRef]

54. Ilhm, L.; Jacobsen, O.; Miiller, R.H.; Mak, E.; Davis, S.S. Surface characteristics and and the interaction of colloidal particles with mouse peritoneal macrophages. *Biomaterials* **1986**, *8*, 113–117.

55. Beyer, D.; Knoll, W.; Ringsdorf, H.; Wang, J.; Timmons, R.B.; Sluka, P. Reduced protein adsorption on plastics via direct plasma deposition of triethylene glycol monoallyl ether. *J. Biomed. Mater. Res.* **1997**, *36*, 181–189. [CrossRef]

56. Favia, P.; Sardella, E.; Gristina, R.; d'Agostino, R. Novel plasma processes for biomaterials: Micro-scale patterning of biomedical polymers. *Surf. Coat. Technol.* **2003**, *169–170*, 707–711. [CrossRef]

57. Krsko, P.; Kaplan, J.B.; Libera, M. Spatially controlled bacterial adhesion using surface-patterned poly(ethylene glycol) hydrogels. *Acta Biomater.* **2009**, *5*, 589–596. [CrossRef] [PubMed]

58. Johnston, E.E.; Bryers, J.D.; Ratner, B.D. Plasma deposition and surface characterization of oligoglyme, dioxane, and crown ether nonfouling films. *Langmuir* **2005**, *21*, 870–881. [CrossRef] [PubMed]

59. Kingshott, P.; Wei, J.; Bagge-Ravn, D.; Gadegaard, N.; Gram, L. Covalent attachment of poly(ethylene glycol) to surfaces, critical for reducing bacterial adhesion. *Langmuir* **2003**, *19*, 6912–6921. [CrossRef]

60. Rodriguez-Emmenegger, C.; Kylián, O.; Houska, M.; Brynda, E.; Artemenko, A.; Kousal, J.; Alles, A.B.; Biederman, H. Substrate-independent approach for the generation of functional protein resistant surfaces. *Biomacromolecules* **2011**, *12*, 1058–1066. [CrossRef] [PubMed]

61. Şen, Y.; Bağci, U.; Güleç, H.A.; Mutlu, M. Modification of Food-Contacting Surfaces by Plasma Polymerization Technique: Reducing the Biofouling of Microorganisms on Stainless Steel Surface. *Food Bioprocess Technol.* **2012**, *5*, 166–175. [CrossRef]

62. Li, M.S.; Zhao, Z.P.; Wang, M.X.; Zhang, Y. Controllable modification of polymer membranes by LDDLT plasma flow: Antibacterial layer onto PE hollow fiber membrane module. *Chem. Eng. J.* **2015**, *265*, 16–26. [CrossRef]

63. Johnston, E.; Bryers, J.E.; Ratner, B.D. Interaction between Pseudomonas aeuginosa and plasma-deposited PEO-like thin films during initial attachment and growth. *Am. Chem. Soc. Polym. Prepr. Division Polym. Chem.* **1997**, *38*, 1016–1017.

64. Stoffels, E. "Tissue processing" with atmospheric plasmas. *Contrib. Plasma Phys.* **2007**, *47*, 40–48. [CrossRef]

65. Da Ponte, G.; Sardella, E.; Fanelli, F.; D'Agostino, R.; Gristina, R.; Favia, P. Plasma deposition of PEO-like coatings with aerosol-assisted dielectric barrier discharges. *Plasma Process. Polym.* **2012**, *9*, 1176–1183. [CrossRef]

66. Da Ponte, G.; Sardella, E.; Fanelli, F.; Van Hoeck, A.; d'Agostino, R.; Paulussen, S.; Favia, P. Atmospheric pressure plasma deposition of organic films of biomedical interest. *Surf. Coat. Technol.* **2011**, *205*, S525–S528. [CrossRef]

67. Mauchauffé, R.; Moreno-Couranjou, M.; Boscher, N.D.; Van De Weerdt, C.; Duwez, A.S.; Choquet, P. Robust bio-inspired antibacterial surfaces based on the covalent binding of peptides on functional atmospheric plasma this films. *J. Mater. Chem. B.* **2014**, *2*, 5168–5177. [CrossRef]

68. Quintieri, L.; Pistillo, B.R.; Caputo, L.; Favia, P.; Baruzzi, F. Bovine lactoferrin and lactoferricin on plasma-deposited coating against spoilage Pseudomonas spp. *Innov. Food Sci. Emerg. Technol.* **2013**, *20*, 215–222. [CrossRef]

69. Camporeale, G.; Moreno-Couranjou, M.; Bonot, S.; Mauchauffé, R.; Boscher, N.D.; Bebrone, C.; Van de Weerdt, C.; Cauchie, H.M.; Favia, P.; Choquet, P. Atmospheric-Pressure Plasma Deposited Epoxy-Rich Thin Films as Platforms for Biomolecule Immobilization-Application for Anti-Biofouling and Xenobiotic-Degrading Surfaces. *Plasma Process. Polym.* **2015**, 1208–1219. [CrossRef]

70. Jampala, S.N.; Sarmadi, M.; Somers, E.B.; Wong, A.C.L.; Denes, F.S. Plasma-enhanced synthesis of bactericidal quaternary ammonium thin layers on stainless steel and cellulose surfaces. *Langmuir* **2008**, *24*, 8583–8591. [CrossRef] [PubMed]

71. Schofield, W.C.E.; Badyal, J.P.S. A substrate-independent approach for bactericidal surfaces. *ACS Appl. Mater. Interfaces* **2009**, *1*, 2763–2767. [CrossRef] [PubMed]

72. Hancock, R.E. Cationic peptides: Effectors in innate immunity and novel antimicrobials. *Lancet Infect. Dis.* **2001**, *1*, 156–164. [CrossRef]

73. Magliulo, M.; Mallardi, A.; Mulla, M.Y.; Cotrone, S.; Pistillo, B.R.; Favia, P.; Vikholm-Lundin, I.; Palazzo, G.; Torsi, L. Electrolyte-gated organic field-effect transistor sensors based on supported biotinylated phospholipid bilayer. *Adv. Mater.* **2013**, *25*, 2090–2094. [CrossRef] [PubMed]

74. De Bartolo, L.; Morelli, S.; Piscioneri, A.; Lopez, L.C.; Favia, P.; d'Agostino, R.; Drioli, E. Novel membranes and surface modification able to activate specific cellular responses. *Biomol. Eng.* **2007**, *24*, 23–26. [CrossRef] [PubMed]

75. Favia, P.; Palumbo, F.; d'Agostino, R.; Lamponi, S.; Magnani, A.; Barbucci, R. Immobilization of Heparin and Highly-Sulphated Hyaluronic Acid onto Plasma-Treated Polyethylene. *Plasmas Polym.* **1998**, *3*, 77–96. [CrossRef]

76. Kastellorizios, M.; Michanetzis, G.P.A.K.; Pistillo, B.R.; Mourtas, S.; Klepetsanis, P.; Favia, P.; Sardella, E.; D'Agostino, R.; Missirlis, Y.F.; Antimisiaris, S.G. Haemocompatibility improvement of metallic surfaces by covalent immobilization of heparin-liposomes. *Int. J. Pharm.* **2012**, *432*, 91–98. [CrossRef] [PubMed]

77. Lopez, L.C.; Gristina, R.; Ceccone, G.; Rossi, F.; Favia, P.; d'Agostino, R. Immobilization of RGD peptides on stable plasma-deposited acrylic acid coatings for biomedical devices. *Surf. Coat. Technol.* **2005**, *200*, 1000–1004. [CrossRef]

78. Bazaka, K.; Jacob, M.V.; Chrzanowski, W.; Ostrikov, K. Anti-bacterial surfaces: Natural agents, mechanisms of action, and plasma surface modification. *RSC Adv.* **2015**, *5*, 48739–48759. [CrossRef]

79. Heyse, P.; Roeffaers, M.B.J.; Paulussen, S.; Hofkens, J.; Jacobs, P.A.; Sels, B.F. Protein immobilization using atmospheric-pressure dielectric-barrier discharges: A route to a straightforward manufacture of bioactive films. *Plasma Process. Polym.* **2008**, *5*, 186–191. [CrossRef]

80. Heyse, P.; Van Hoeck, A.; Roeffaers, M.B.J.; Raffin, J.P.; Steinbüchel, A.; Stöveken, T.; Lammertyn, J.; Verboven, P.; Jacobs, P.A.; Hofkens, J.; et al. Exploration of atmospheric pressure plasma nanofilm technology for straightforward bio-active coating deposition: Enzymes, plasmas and polymers, an elegant synergy. *Plasma Process. Polym.* **2011**, *8*, 965–974. [CrossRef]

81. Da Ponte, G.; Sardella, E.; Fanelli, F.; Paulussen, S.; Favia, P. Atmospheric pressure plasma deposition of poly lactic acid-like coatings with embedded elastin. *Plasma Process. Polym.* **2014**, *11*, 345–352. [CrossRef]

82. Palumbo, F.; Camporeale, G.; Yang, Y.W.; Wu, J.S.; Sardella, E.; Dilecce, G.; Calvano, C.D.; Quintieri, L.; Caputo, L.; Baruzzi, F.; et al. Direct Plasma Deposition of Lysozyme-Embedded Bio-Composite Thin Films. *Plasma Process. Polym.* **2015**, *12*, 1302–1310. [CrossRef]

83. Yang, Y.W.; Camporeale, G.; Sardella, E.; Dilecce, G.; Wu, J.S.; Palumbo, F.; Favia, P. Deposition of hydroxyl functionalized films by means of ethylene aerosol-assisted atmospheric pressure plasma. *Plasma Process. Polym.* **2014**, *11*, 1102–1111. [CrossRef]

84. Palza, H. Antimicrobial polymers with metal nanoparticles. *Int. J. Mol. Sci.* **2015**, *16*, 2099–2116. [CrossRef] [PubMed]

85. Sintubin, L.; De Windt, W.; Dick, J.; Mast, J.; Van Der Ha, D.; Verstraete, W.; Boon, N. Lactic acid bacteria as reducing and capping agent for the fast and efficient production of silver nanoparticles. *Appl. Microbiol. Biotechnol.* **2009**, *84*, 741–749. [CrossRef] [PubMed]

86. Grass, G.; Rensing, C.; Solioz, M. Metallic copper as an antimicrobial surface. *Appl. Environ. Microbiol.* **2011**, *77*, 1541–1547. [CrossRef] [PubMed]

87. Zhang, Y.M.; Rock, C.O. Membrane lipid homeostasis in bacteria. *Nat. Rev. Microbiol.* **2008**, *6*, 222–233. [CrossRef] [PubMed]

88. Mathews, S.; Hans, M.; Mücklich, F.; Solioz, M. Contact killing of bacteria on copper is suppressed if bacterial-metal contact is prevented and is induced on iron by copper ions. *Appl. Environ. Microbiol.* **2013**, *79*, 2605–2611. [CrossRef] [PubMed]

89. Mulley, G.; Jenkins, A.T.A.; Waterfield, N.R. Inactivation of the antibacterial and cytotoxic properties of silver ions by biologically relevant compounds. *PLoS ONE* **2014**, *9*, 2–10. [CrossRef] [PubMed]

90. Reidy, B.; Haase, A.; Luch, A.; Dawson, K.A.; Lynch, I. Mechanisms of silver nanoparticle release, transformation and toxicity: A critical review of current knowledge and recommendations for future studies and applications. *Materials* **2013**, *6*, 2295–2350. [CrossRef]

91. Bayston, R.; Vera, L.; Mills, A.; Ashraf, W.; Stevenson, O.; Howdle, S.M. In vitro antimicrobial activity of silver-processed catheters for neurosurgery. *J. Antimicrob. Chemother.* **2009**, *65*, 258–265. [CrossRef] [PubMed]

92. Kvitek, L.; Panacek, A.; Prucek, R.; Soukupova, J.; Vanickova, M.; Kolar, M.; Zboril, R. Antibacterial activity and toxicity of silver—Nanosilver versus ionic silver. *J. Phys. Conf. Ser.* **2011**, *304*, 012029. [CrossRef]

93. Bruellhoff, K.; Fiedler, J.; Möller, M.; Groll, J.; Brenner, R.E. Surface coating strategies to prevent biofilm formation on implant surfaces. *Int. J. Artif. Organs* **2010**, *33*, 646–653. [PubMed]

94. Deng, X.; Nikiforov, A.Y.; Coenye, T.; Cools, P.; Aziz, G.; Morent, R.; De Geyter, N.; Leys, C. Antimicrobial nano-silver non-woven polyethylene terephthalate fabric via an atmospheric pressure plasma deposition process. *Sci. Rep.* **2015**, *5*, 10138. [CrossRef] [PubMed]

95. Deng, X.; Nikiforov, A.; Vujosevic, D.; Vuksanovic, V.; Mugoša, B.; Cvelbar, U.; De Geyter, N.; Morent, R.; Leys, C. Antibacterial activity of nano-silver non-woven fabric prepared by atmospheric pressure plasma deposition. *Mater. Lett.* **2015**, *149*, 95–99. [CrossRef]

96. Zille, A.; Fernandes, M.M.; Francesko, A.; Tzanov, T.; Fernandes, M.; Oliveira, F.R.; Almeida, L.; Amorim, T.; Carneiro, N.; Esteves, M.F.; et al. Size and Aging Effects on Antimicrobial Efficiency of Silver Nanoparticles Coated on Polyamide Fabrics Activated by Atmospheric DBD Plasma. *ACS Appl. Mater. Interfaces* **2015**, *7*, 13731–13744. [CrossRef] [PubMed]

97. Vasilev, K.; Sah, V.; Anselme, K.; Ndi, C.; Mateescu, M.; Dollmann, B.; Martinek, P.; Ys, H.; Ploux, L.; Griesser, H.J. Tunable antibacterial coatings that support mammalian cell growth. *Nano Lett.* **2010**, *10*, 202–207. [CrossRef] [PubMed]

98. Ploux, L.; Mateescu, M.; Anselme, K.; Vasilev, K. Antibacterial properties of silver-loaded plasma polymer coatings. *J. Nanomater.* **2012**, *2012*. [CrossRef]

99. Kumar, V.; Jolivalt, C.; Pulpytel, J.; Jafari, R.; Arefi-Khonsari, F. Development of silver nanoparticle loaded antibacterial polymer mesh using plasma polymerization process. *J. Biomed. Mater. Res. A* **2013**, *101A*, 1121–1132. [CrossRef] [PubMed]

100. Pathak, S.C.; Hess, D.W. Dissolution and swelling behaviour of plasma-polymerized polyethylene glycol-like hydrogel films for use as drug-delivery reservoir. *ECS Trans.* **2008**, *6*, 1–12.

101. Balazs, D.J.; Triandafillu, K.; Sardella, E.; Iacoviello, G.; Favia, P.; d'Agostino, R.; Harms, H.; Mathieu, H.J. PECVD modification of PVC endotracheal tubes to inhibit bacterial adhesion: PEO-like and nano-composite Ag/PEO-like coatings. In *Plasma Processes and Polymers*; d'Agostino, R., Favia, P., Oher, C., Wertheimer, M.R., Eds.; Wiley VCH: Weinheim, Germany, 2005; pp. 351–372.

102. Liu, X.; Zhang, C.; Yang, J.; Lin, D.; Zhang, L.; Chen, X.; Zha, L. Silver nanoparticles loading pH responsive hybrid microgels: pH tunable plasmonic coupling demonstrated by surface enhanced Raman scattering. *RSC Adv.* **2013**, *3*, 3384. [CrossRef]

103. Song, W.-H.; Ryu, H.S.; Hong, S.-H. Antibacterial properties of Ag (or Pt)-containing calcium phosphate coatings formed by micro-arc oxidation. *J. Biomed. Mater. Res. A* **2009**, *88*, 246–254. [CrossRef] [PubMed]

104. Hu, H.; Zhang, W.; Qiao, Y.; Jiang, X.; Liu, X.; Ding, C. Antibacterial activity and increased bone marrow stem cell functions of Zn-incorporated TiO$_2$ coatings on titanium. *Acta Biomater.* **2012**, *8*, 904–915. [CrossRef] [PubMed]

105. Duque, L.; Förch, R. Plasma polymerization of zinc acetyl acetonate for the development of a polymer-based zinc release system. *Plasma Process. Polym.* **2011**, *8*, 444–451. [CrossRef]

106. Poulter, N.; Munoz-Berbel, X.; Johnson, A.L.; Dowling, A.J.; Waterfield, N.; Jenkins, A.T.A. An organo-silver compound that shows antimicrobial activity against Pseudomonas aeruginosa as a monomer and plasma deposited film. *Chem. Commun. (Camb.)* **2009**, *2009*, 7312–7314. [CrossRef] [PubMed]

107. Liguori, A.; Traldi, E.; Toccaceli, E.; Laurita, R.; Pollicino, A.; Focarete, M.L.; Colombo, V.; Gherardi, M. Co-Deposition of Plasma-Polymerized Polyacrylic Acid and Silver Nanoparticles for the Production of Nanocomposite Coatings Using a Non-Equilibrium Atmospheric Pressure Plasma Jet. *Plasma Process. Polym.* **2015**, *13*, 623–632. [CrossRef]

108. Beier, O.; Pfuch, A.; Horn, K.; Weisser, J.; Schnabelrauch, M.; Schimanski, A. Low temperature deposition of antibacterially active silicon oxide layers containing silver nanoparticles, prepared by atmospheric pressure plasma chemical vapor deposition. *Plasma Process. Polym.* **2013**, *10*, 77–87. [CrossRef]

109. Deng, X.; Leys, C.; Vujosevic, D.; Vuksanovic, V.; Cvelbar, U.; De Geyter, N.; Morent, R.; Nikiforov, A. Engineering of Composite Organosilicon Thin Films with Embedded Silver Nanoparticles via Atmospheric Pressure Plasma Process for Antibacterial Activity. *Plasma Process. Polym.* **2014**, *11*, 921–930. [CrossRef]

110. Choi, H.W.; Dauskardt, R.H.; Lee, S.-C.; Lee, K.-R.; Oh, K.H. Characteristic of silver doped DLC films on surface properties and protein adsorption. *Diam. Relat. Mater.* **2008**, *17*, 252–257. [CrossRef]

111. Baba, K.; Hatada, R.; Flege, S.; Ensinger, W.; Shibata, Y.; Nakashima, J.; Sawase, T.; Morimura, T. Preparation and antibacterial properties of Ag-containing diamond-like carbon films prepared by a combination of magnetron sputtering and plasma source ion implantation. *Vacuum* **2013**, *89*, 179–184. [CrossRef]

112. Saulou, C.; Despax, B.; Raynaud, P.; Zanna, S.; Seyeux, A.; Marcus, P.; Audinot, J.N.; Mercier-Bonin, M. Plasma-mediated nanosilver-organosilicon composite films deposited on stainless steel: Synthesis, surface characterization, and evaluation of anti-adhesive and anti-microbial properties on the model yeast saccharomyces cerevisiae. *Plasma Process. Polym.* **2012**, *9*, 324–338. [CrossRef]

113. Körner, E.; Aguirre, M.H.; Fortunato, G.; Ritter, A.; Rühe, J.; Hegemann, D. Formation and distribution of silver nanoparticles in a functional plasma polymer matrix and related Ag$^+$ release properties. *Plasma Process. Polym.* **2010**, *7*, 619–625. [CrossRef]

114. Chakravadhanula, V.S.K.; Kübel, C.; Hrkac, T.; Zaporojtchenko, V.; Strunskus, T.; Faupel, F.; Kienle, L. Surface segregation in TiO$_2$-based nanocomposite thin films. *Nanotechnology* **2012**, *23*, 495701. [CrossRef] [PubMed]

115. **Drábik, M.;** Pešička, J.; Biederman, H.; Hegemann, D. Long-term aging of Ag/a-C:H:O nanocomposite coatings in air and in aqueous environment. *Sci. Technol. Adv. Mater.* **2015**, *16*, 025005. [CrossRef]

116. Escobar Galindo, R.; Manninen, N.K.; Palacio, C.; Carvalho, S. Advanced surface characterization of silver nanocluster segregation in Ag-TiCN bioactive coatings by RBS, GDOES, and ARXPS. *Anal. Bioanal. Chem.* **2013**, *405*, 6259–6269. [CrossRef] [PubMed]

117. Katsikogianni, M.G.; Foka, A.; Sardella, E.; Ingrosso, C.; Favia, P.; Mangone, A.; Spiliopoulou, I.; Missirlis, Y.F. Fluid Flow and Sub-Bactericidal Release of Silver from Organic Nanocomposite Coatings Enhance *ica* Operon Expression in *Staphylococcus epidermidis*. *Sci. Res.* **2013**, *4*, 30–40.

118. Kajiyama, S.; Tsurumoto, T.; Osaki, M.; Yanagihara, K.; Shindo, H. Quantitative analysis of *Staphylococcus epidermidis* biofilm on the surface of biomaterial. *J. Orthop. Sci.* **2009**, *14*, 769–775. [CrossRef] [PubMed]

119. D'Agostino, R.; Favia, P.; Fracassi, F.; Sardella, E.; Costigliola, A.; Mangone, A. Processes for the Production by Plasma of Nanometric Thickness Coatings Allowing Controlled Release of Silver Ions of Other Elements, or of Molecules of Biomedical Interest, from Solid Products, and Products thus Coated. Patent WO 2013021409 A8, 4 April 2013.

120. Kvítek, L.; Panácek, A.; Soukupova, J.; Kolar, M.; Vecerova, R.; Prucek, R.; Holecova, M.; Zboril, R.; Kvı, L.; Vec, R.; et al. Effect of surfactants and polymers on stability and antibacterial activity of silver nanoparticles. *J. Phys. Chem.* **2008**, *112*, 5825–5834. [CrossRef]

121. Laroussi, M. From killing bacteria to destroying cancer cells: 20 years of plasma medicine. *Plasma Process. Polym.* **2014**, *11*, 1138–1141. [CrossRef]

122. Randeniya, L.K.; De Groot, G.J.J.B. Non-Thermal Plasma Treatment of Agricultural Seeds for Stimulation of Germination, Removal of Surface Contamination and Other Benefits: A Review. *Plasma Process. Polym.* **2015**, *12*, 608–623. [CrossRef]

materials

MDPI

Article

Investigation of Industrial Polyurethane Foams Modified with Antimicrobial Copper Nanoparticles

Maria Chiara Sportelli [1], Rosaria Anna Picca [1], Roberto Ronco [1], Elisabetta Bonerba [2], Giuseppina Tantillo [2], Mauro Pollini [3], Alessandro Sannino [3], Antonio Valentini [4], Tommaso R.I. Cataldi [1] and Nicola Cioffi [1,*

[1] Dipartimento di Chimica, Università Degli Studi di Bari "Aldo Moro", via E. Orabona 4, Bari (BA) 70126, Italy; maria.sportelli@uniba.it (M.C.S.); rosaria.picca@uniba.it (R.A.P.); robertoronco1@libero.it (R.R.); tommaso.cataldi@uniba.it (T.R.I.C.)
[2] Dipartimento di Medicina Veterinaria, Università Degli Studi di Bari "Aldo Moro", St.da P.le per Casamassima Km 3, Valenzano (BA) 70010, Italy; elisabetta.bonerba@uniba.it (E.B.); giuseppina.tantillo@uniba.it (G.T.)
[3] Dipartimento di Ingegneria dell'Innovazione, Università del Salento, via per Monteroni, Lecce (LE) 73100, Italy; mauro.pollini@unisalento.it (M.P.); alessandro.sannino@unisalento.it (A.S.)
[4] Dipartimento Interateneo di Fisica, Università Degli Studi di Bari "Aldo Moro", via Amendola 173, Bari (BA) 70126, Italy; antonio.valentini@uniba.it
* Correspondence: nicola.cioffi@uniba.it; Tel.: +39-080-544-2020

Academic Editor: Carla Renata Arciola
Received: 22 April 2016; Accepted: 29 June 2016; Published: 7 July 2016

Abstract: Antimicrobial copper nanoparticles (CuNPs) were electrosynthetized and applied to the controlled impregnation of industrial polyurethane foams used as padding in the textile production or as filters for air conditioning systems. CuNP-modified materials were investigated and characterized morphologically and spectroscopically, by means of Transmission Electron Microscopy (TEM), and X-ray Photoelectron Spectroscopy (XPS). The release of copper ions in solution was studied by Electro-Thermal Atomic Absorption Spectroscopy (ETAAS). Finally, the antimicrobial activity of freshly prepared, as well as aged samples—stored for two months—was demonstrated towards different target microorganisms.

Keywords: copper nanoparticle; polyurethane foam; ETAAS; XPS; nanoantimicrobials

1. Introduction

Application of nanotechnology in the textile industry is practiced with two main purposes. The first one aims at the improvement of the textile performance: fibers modified by nanopowders, or carbon nanotubes (CNTs), generally possess higher mechanical resistance in respect with untreated ones [1]. The second purpose deals with the development of multifunctional textiles, i.e., antimicrobial [2], anti-stain [3], water repellent [4,5], anti-static [6–8], and self-cleaning ones [9]. Cotton, linen, silk and wool are well-known natural fibers and padding materials, breathable and resistant [10]. However, they are easily degraded by many microorganisms, like spores, molds and bacteria [11,12]. Special finishes, however, can provide them protection from biodegradation: nanotechnology offers a good range of effective tools to protect fabrics from bacterial contamination [2]. During the last century, the discovery and the growing production of polymer-based artificial fibers and paddings completely eliminated the problems related to biodegradation. Polyurethane foams play a fundamental role in the production of seats in the automotive sector, of filters for the conditioning and treatment of air and water, and so on. However, in all these fields, prevention of bacterial adhesion and/or proliferation is extremely important. Metal and metal oxide nanoparticles (NPs), specifically Cu and its oxides, have been already used for this purpose [13,14]. Natural fiber modification with copper is rather prevalent and different reports are present in literature on this topic, especially for the treatment of cotton fabrics [14]. Most of them deal with their CuO or Cu_2O functionalization [15–23], while the use of elemental Cu is reported less frequently [24,25]. Cu(0), in fact, easily undergoes oxidation processes unless suitable stabilizing agents are used. Nevertheless, modification of artificial polymers by nanostructured Cu is less diffused, and it is generally achieved by adding a copper-containing additive to the polymers during the master batch preparation stage [26–29]. Silicone fibers [30], polyester [26,31,32], and nylon [33] have been successfully modified by both nanostructured CuO and Cu. Polyurethane (PU) has been modified by different types of inorganic clusters, such as Ag [34–44], CNTs (carbon nanotubes) [45], Zn-Ag bimetallic particles [46], tourmaline [47,48], silica [49] and ZnO [50]. Ag is nowadays one of the preferred additives to confer antimicrobial properties to both natural and synthetic fibers, being well known its antiseptic effect against a wide range of microorganisms. Ag-modified PU foams or fibers are fairly common: in fact, silver is able to confer antimicrobial properties and to improve polymer mechanical and rheological properties [43]. However, to the best of our knowledge, only a few reports are present in the literature about the modification of polyurethanes with antimicrobial CuNPs [51–53].

Here we report on polyurethane foams modified with colloidal CuNPs that were electrochemically synthesized by means of the so-called sacrificial anode electrolysis (SAE) technique [54]. Post-production functionalization of industrial polyurethane foams was carried out by their impregnation in diluted CuNP colloids. Samples were morphologically and spectroscopically investigated and characterized. Their antimicrobial activity was tested towards three model microorganisms (*Staphylococcus aureus*, *Escherichia coli* and *Kluyveromyces marxianus*), demonstrating CuNPs capability of strongly inhibiting bacterial growth and proliferation.

2. Results and Discussion

2.1. Synthesis and Characterization of CuNPs

CuNPs, used as additives of industrial polyurethane foams, were prepared by sacrificial anode electrolysis as described in the experimental section. Cu concentration in the stock colloidal solution was obtained by differential weighing of working and counter electrodes: it resulted equal to 0.31 ± 0.05 M in as-synthesized Cu-nanocolloids. TOAC (tetraoctylammonium chloride) was chosen as stabilizing agent, due to the high reaction yield of TOAC-driven syntheses ($85\% \pm 5\%$), excellent morphological control, and lasting storage times of the colloid up to months; there is only the need to avoid air exposure and thermal shocks. Moreover, TOAC octyl chains confer lipophilicity to metal NPs, which become scarcely soluble in water [55]. This attribute can be proficiently used to promote PU impregnation with CuNPs. With regard to pristine Cu-nanocolloids, the high degree of stabilization provided by TOAC capping agent was demonstrated by the TEM micrograph reported in Figure 1a. The electrochemical process resulted in a NP population with a single mode and narrow diameter dispersion, centered at 2.6 nm ($\sigma \leqslant 0.5$ nm), as shown in the size distribution histogram of Figure 1b.

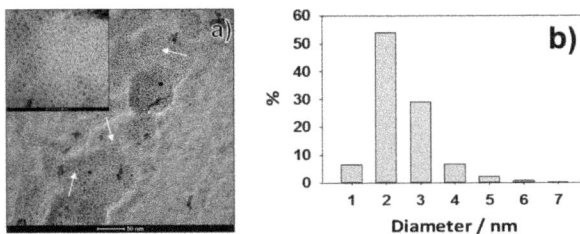

Figure 1. (**a**) Transmission Electron Microscopy (TEM) micrographs of CuNPs (highlighted by arrows) synthesized by sacrificial anode electrolysis. A micrograph at higher magnification is reported as insert; (**b**) Size distribution histogram of as synthesized CuNPs.

2.2. Adsorption of Industrial Polyurethane Foams with CuNPs

PU samples were treated with CuNP-colloids as reported in Section 3.3. Two different types of polyurethane foam were used. For clarity of presentation, they were labeled as follows:

- Sample A: green foam, with large and irregular pores, used as filling material for mattresses (density: 25 kg/m^3, density tolerance ±5%);
- Sample B: white foam, with small and regular pores, used in the automotive industry (density: 21 kg/m^3, density tolerance ±5%).

Adsorption efficiency was tested by weighing each sample before and after the modification process. Both samples showed an average weight increase equal to 8.2 ± 0.3 mg for type A PU and 7.8 ± 0.2 mg for type B PU, respectively. Pristine and treated foams were subjected to optical microscopy examination. This aimed at ascertaining that the impregnation process did not alter PU pores morphology and/or dimension. Typical optical micrographs are reported in Figure 2. A bigger and more dispersed pore size was observed for pristine samples A (Figure 2a,c,e), while pristine samples B were characterized by a more regular porous structure (Figure 2b,d,f). These features remained unaffected after the impregnation process (Figure 2g,h).

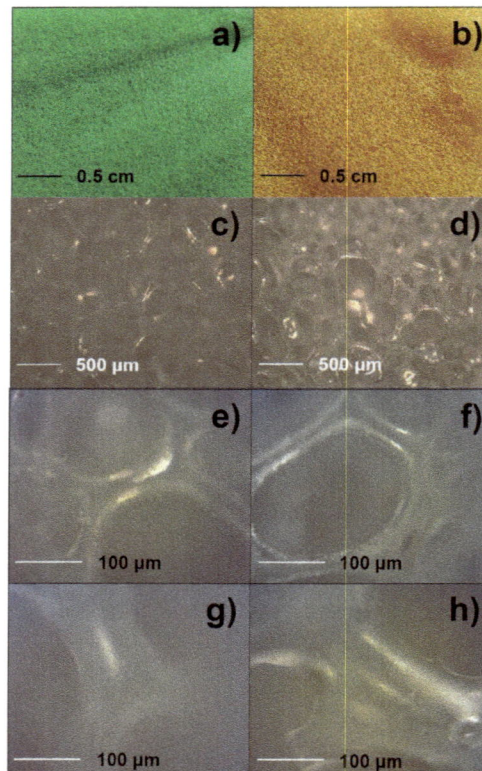

Figure 2. Photographs of (**a**) type A and (**b**) type B pristine polyurethane (PU) foams. Optical micrographs of pristine and Cu-treated foams; (**c**) Low magnification—Pristine type A foam; (**d**) Low magnification—Pristine type B foam; (**e**) High magnification—Pristine type A foam; (**f**) High magnification—Pristine type B foam; (**g**) 1:1000 Cu-modified type A foam; (**h**) 1:1000 Cu-modified type B foam.

2.3. Surface Chemical Characterization of CuNP-Modified Industrial Polyurethane Foams

In order to assess the surface chemical composition of CuNP-modified polyurethane foams, XPS (X-ray photoelectron spectroscopy) analyses were performed. Both A and B PU samples were studied before and after treatment with CuNPs colloids, purposely suspension dilute 100- and 1000-fold with pure solvent. Data of Table 1 show that carbon and oxygen were the most abundant elements on the analyzed surfaces, as expected for an oxygenated polymer dispersing matrix modified by low amounts of CuNPs.

Table 1. Surface elemental composition estimated by X-ray photoelectron spectroscopy (XPS) of fresh samples A and B, treated with CuNPs. Error is expressed as the larger value between the error associated to a single quantification (0.2% for copper, 0.5% for other elements) and one standard deviation, calculated on at least three replicate analyses. Data about pristine samples are reported for comparison.

Element	Sample A			Sample B		
	Pristine	PU/CuNPs (1:100)	PU/CuNPs (1:1000)	Pristine	PU/CuNPs (1:100)	PU/CuNPs (1:1000)
Cu	<0.2%	1.3 ± 0.2	0.8 ± 0.2	<0.2%	0.5 ± 0.2	0.3 ± 0.2
C	73.7 ± 0.5	76.9 ± 0.5	79 ± 3	72.6 ± 0.5	68.4 ± 0.5	67.5 ± 0.5
N	1.6 ± 0.5	1.7 ± 0.5	1.3 ± 0.5	1.6 ± 0.5	1.6 ± 0.5	2.8 ± 0.5
O	23.3 ± 0.5	18.5 ± 0.5	18 ± 3	20.7 ± 0.5	24.4 ± 0.5	23.7 ± 0.5
Si	1.4 ± 0.5	1.6 ± 0.5	0.9 ± 0.5	4.5 ± 0.5	5.1 ± 0.5	5.7 ± 0.5
Cl	–	<0.5	<0.5	–	<0.5	<0.5
Ca	–	–	–	0.6 ± 0.5	<0.5	<0.5

Cu traces could be qualitatively observed on pristine samples, although they were below the XPS limit of quantification. This evidence is reasonable, considering that extrusion of PU foams is performed using copper dyes. Small amounts of Si were also identified: this element is characteristic of PU and PU-based goods production processes [56]. Presence of nitrogen could be attributed to both TOAC (when present) and polymeric backbone. It is also important to point up how Cu-modified samples A had in general a higher Cu atomic %, compared to homologous composites B (e.g., composites obtained by impregnation with the same CuNP dilution). This might be due to the presence of larger pores, which allowed CuNPs to better penetrate into the bulk material. Chlorine is due to the presence of CuNP stabilizing agent (i.e., TOAC), while traces of calcium observed in samples B were considered as a contamination coming from PU industrial production processes.

XP high-resolution (HR) regions were also investigated, in order to obtain detailed information about PU and CuNPs chemical speciation. HR XP C1s spectra of pristine and treated PU foams were made of three main components. As an example, Table 2 resumes data obtained on samples A in terms of peak position (expressed as Binding Energy—BE—values), relative abundance, and signal attribution.

Table 2. Attributions of C1s chemical environments identified on type A pristine and Cu-modified PU foams; relative abundance % of each signal component is reported for comparison. Error is expressed as one standard deviation, calculated on at least three replicate analyses.

Sample	BE (eV)	Attribution	Relative Abundance %
Pristine	284.8 ± 0.1	C–C	43 ± 2
	286.4 ± 0.2	C–O, C–N	55.8 ± 1.3
	289.0 ± 0.2	HN–C=O	1.2 ± 0.8
PU/CuNPs (1:100)	284.8 ± 0.1	C–C	55 ± 3
	286.4 ± 0.2	C–O, C–N	42 ± 2
	288.8 ± 0.2	HN–C=O	3.0 ± 1.3

All components are in agreement with those expected for PU [57,58]. On Cu-modified foams, a higher relative percentage of aliphatic C–C moieties was observed. This could be explained considering the presence of octyl- alkyl chains from TOAC. HR XP N1s spectra were characterized by a single component in all cases, centered at 400.0 ± 0.2 eV, compatible with PU urethane functionalities [58]. Typical $Cu2p_{3/2}$ XP spectra for fresh and aged samples are reported in Figure 3. Soon after sample preparation (Figure 3a), a single signal falling at 933.2 ± 0.2 eV was present. This peak can be ascribed both to nanodispersed Cu at zero oxidation state [58–60] and/or to cuprous species [57,58]. The absence of Cu(II) moieties was confirmed by the lack of photoelectronic signals in the region centered at about 934.0 ± 0.2 eV and the shake-up features between 941.0 and 948.0 eV, as well [57]. CuNPs undergo a certain surface oxidation, as a function of the aging time. This phenomenon led to the formation of cupric moieties onto modified PU foams. $Cu2p_{3/2}$ XP high-resolution region showed, for aged samples (Figure 3b), the presence of a second signal component, at higher binding energy values (934.4 ± 0.3 eV), ascribable to Cu(II) moieties. Shake-up bands were, in this case, barely visible, due to the very high signal-to-noise ratio.

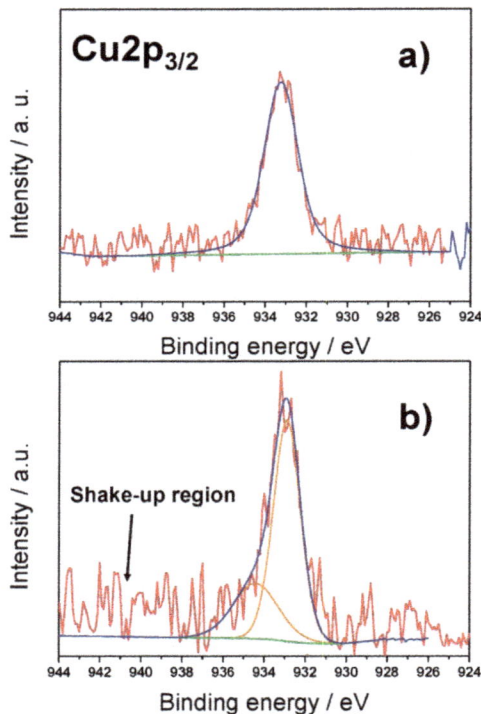

Figure 3. Typical $Cu2p_{3/2}$ X-ray photoelectron (XP) high-resolution spectra of fresh (**a**); and aged (**b**) CuNP-modified polyurethane foams.

2.4. Kinetics of Copper Release from CuNP-Modified Industrial Polyurethane Foams

CuNP-modified PUs were exposed to a physiological saline solution in order to mimic a possible interaction and ionic release of nanocomposites to model contact media. Experiments were carried out on both freshly prepared and aged samples, which were stored in air for 60 days. The kinetics of Cu release, relative to all the analyzed samples are shown in Figure 4. The experimental data could be interpolated, in all cases, by a pseudo-first order kinetic model.

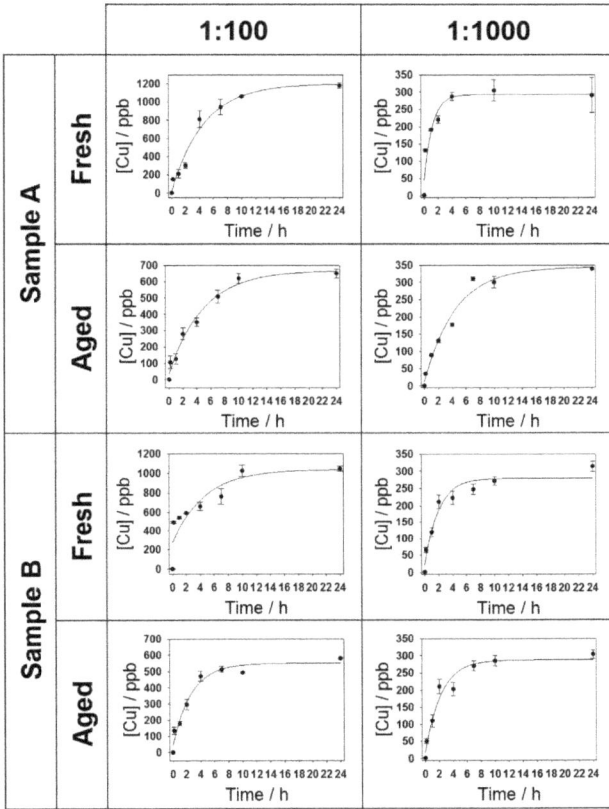

Figure 4. Copper release in physiologic solution from CuNP-modified polyurethane foams, as a function of the incubation time. Different columns are relevant to different CuNP concentrations in the impregnation baths, giving rise to different Cu surface abundance on the composite. Different rows are relevant to composites obtained by different polyurethane batches or to differently aged samples.

From the analysis of all the curves, it was found that part of the Cu ion release in solution was quite fast in the first ten minutes. This is reasonably due to the presence of readily soluble Cu species (such as residual cuprous salts) on the surface of CuNPs. It was also evident, in all cases, that Cu concentration progressively grew as a function of time up to a plateau value. The extent of the latter varied according to the concentration of the impregnation bath, and as a function of the aging time, too. In both samples, in fact, a higher mean plateau value was registered for PU foams treated with 1:100 colloidal dilutions, which had the higher surface Cu concentration. The larger pore size of composites based on A-type PU might explain the slightly higher and faster Cu release recorded for materials treated with 1:100 CuNPs dilution, due to a possible easier accessibility of the physiological solution within samples A. In principle, a larger pore size should allow a deeper and easier penetration of physiological solution within the foam. Moreover, for both samples A and B, aging resulted in a considerable decrease of the Cu release. A tentative interpretation of this phenomenon was that surfactant could, over time, segregate [61–63] and partially occlude PU pores, enhancing the surface hydrophobicity of the composite and partially limiting the accessibility of the physiological solution within them. The average values of plateau Cu concentrations, $[Cu]^0$, and kinetic constants for all samples are summarized in Table 3.

Table 3. Average values of plateau Cu concentrations and kinetic constants for samples A and B. Data about fresh and aged samples are reported. The error is expressed as the largest value between the standard deviation relevant to the repeated measurements and the error associated to individual quantifications.

Sample	CuNPs Dilution	Plateau [Cu]/ppb		Kinetic Constant/h^{-1}		[Cu]^0/ppb	
		Fresh	Aged	Fresh	Aged	Fresh	Aged
Sample A	1:100	1200 ± 90	670 ± 40	0.20 ± 0.04	0.20 ± 0.04	0	40 ± 30
	1:1000	300 ± 40	330 ± 20	0.8 ± 0.3	0.20 ± 0.04	50 ± 30	0
Sample B	1:100	1100 ± 200	550 ± 40	0.2 ± 0.1	0.40 ± 0.09	300 ± 100	40 ± 30
	1:1000	260 ± 30	270 ± 20	0.5 ± 0.1	0.4 ± 0.1	30 ± 20	20 ± 20

2.5. Antimicrobial Tests

The antimicrobial properties of freshly prepared Cu-polyurethane composites were evaluated in preliminary tests on three target microorganisms selected based on microorganisms characteristics (Gram-positive, Gram-negative, yeasts) to demonstrate the broad-spectrum of NP antimicrobial activity. The suitable dilution tested was obtained taking into account the pathogenic role of microorganisms (mainly *S. aureus* and *E. coli*), the spread and persistence of *S. aureus* on the surfaces, the assessment of *E. coli* as hygiene requirement, and the environment ubiquity of *K. marxianus*. Culture broths were diluted by different factors, based on the different characteristics of the microorganisms (10^7 for *S. aureus*, 10^5 for *E. coli* and 10^3 for *K. marxianus*), and then they were left in contact for fixed times with different samples in order to discriminate the biocidal/biostatic effects of pristine PU foams, electrolytic solution, and CuNP-treated materials. In any sample, after a contact time of 24 h, the residual microorganism growth was quantified by counting the number of colony forming units (CFU). The results are reported in Table 4. As expected, blank experiments on pristine PU foams did not show any biostatic action, whereas control samples treated with 0.1 M TOAC solution in Acetonitrile/tetrahydrofuran (ACN/THF) $1/3_{v/v}$ mixture produced a strong inhibition effect against *S. aureus*, causing a complete inhibition of the bacterial growth. This is a reasonable result, since TOAC salt belongs to the class of common quaternary ammonium disinfectants. Using Cu-modified polyurethanes induced a marked growth inhibition even in the case of microorganisms such as *E. coli*, that did not show any sensitivity to TOAC-impregnated PU. Finally, comparing the activity of composites treated by colloidal dilutions of 1:100 and 1:1000, it can be concluded that a higher CuNP loading was generally correlated to a higher concentration of released ions, hence in an increased inhibition of colony growth. In some experiments apparent discrepancies between ionic release and bioactivity were observed and attributed to limited reproducibility of the investigated biological systems.

Table 4. Number of colony forming units (CFU) for the three target microorganisms, exposed to different samples for 24 h as described in the experimental section. Error on CFU counts is ± 5 in the last digit.

	Sample	*S. aureus*/CFU	*E. coli*/CFU	*K. marxianus*/CFU
Sample A	PU	U[a]	U	U
	PU + 0.1 M TOAC solution	0	U	U
	PU + 1:1000 CuNPs	0	0	30
	PU + 1:100 CuNPs	0	0	25
Sample B	PU	U	U	U
	PU + 0.1 M TOAC solution	2	U	U
	PU + 1:1000 CuNPs	U	72	U
	PU + 1:100 CuNPs	0	0	U

[a] U = Uncountable.

3. Materials and Methods

3.1. Materials

Copper (0.5 mm thick, 99.99+%) and platinum sheets (0.25 mm thick, 99.999%) were purchased from Goodfellow Ltd. (Cambridge, UK) and cut into 2 × 1 cm pieces. Acetonitrile (ACN, anhydrous, 99.8%), tetrahydrofuran (THF, anhydrous, ⩾99.9%, inhibitor-free), and tetraoctylammonium chloride (TOAC, AT reagent, ⩾97.0%), were purchased from Sigma Aldrich (Milan, Italy). Aluminum Oxide (purum p.a., 99.7%), for the mechanical polishing of metallic sheets, was from Fluka Chemicals (Milan, Italy). Polyurethane foams were obtained from the industrial partner ME.RES. *Meridionale Resine* S.r.L (Avellino, Italy).

Escherichia coli ATCC 25922, *Staphylococcus aureus* FDA 209P (MSSA, with methicillin resistance 0.125 μg·mL^{-1}), and *Kluyveromyces marxianus* CBS 608, selected as target microorganisms for biological tests, were obtained from BioMérieux Italia S.p.A. (Florence, Italy), and reconstituted in nutrient broth (Agar Oxoid), purchased from Bio-Chemia (Bari, Italy), as the Plate Count Agar culture medium.

3.2. Electrochemical Synthesis of CuNPs

The electrochemical synthesis of CuNPs was carried out in a three-electrode cell equipped with a Cu working electrode and an Ag/AgNO$_3$ 0.1 M in ACN reference electrode. The counter electrode was a Pt sheet. The electrolytic solution was composed of 0.1 M TOAC dissolved in an ACN/THF 1/3 mixture. The electrosynthesis was carried out in nitrogen atmosphere, potentiostatically, (+1.5 V vs. reference) for 6 h, under vigorous stirring at room temperature, with a CH1140b potentiostat-galvanostat (CH Instruments, Austin, TX, USA). Other electrochemical parameters such as electrodes pretreatment, electrolysis cell, process details, etc. were similar to what reported in previous works [61–63].

3.3. Modification of Industrial Polyurethane Foams with CuNPs

Cubic samples, of approximately 1 cm^3, and weighing about 100 mg, were obtained cutting the as-received materials.

After ascertaining the resistance of each sample towards the solvents used for the electrosynthesis, each material was treated with CuNPs by immersion in 5 mL of diluted colloids for 30 min. After this time, each sample was wrung out and left to dry in air for 2 h. Typically, dilution ratios of CuNP stock solution equal to 1:100 and 1:1000 were used.

3.4. Morphological and Spectroscopic Characterization

Colloidal CuNPs diluted in the ratio 1:10 were sonicated for 30 min, before Transmission Electron Microscopy (TEM) analysis, in order to prevent possible aggregation. TEM microscopy was performed with a FEI Tecnai 12 instrument (Hillsboro, OR, USA, high tension: 120 kV; filament: W), by dropping 10 μL on carbon-coated Cu grids (300 mesh, TAAB Laboratories Equipment Ltd., Aldermaston, UK). The microscope was calibrated using the S106 Cross Grating (2160 lines/mm, 3.05 mm) supplied by Agar Scientific (Stansted, UK). Alignment was checked by using factory settings and routines. Astigmatism was adjusted by means of fast Fourier transform processing. Size distribution of metal clusters was evaluated using ImageJ software [64]. Treated and pristine polyurethane materials were characterized by means of X-ray photoelectron spectroscopy (XPS), using a Thermo Fisher Scientific Theta Probe Spectrometer (Waltham, MA, USA). A monochromatized AlKα source was used, with a beam spot diameter of 300 μm. Samples were mounted onto the sample holder by means of a carbon double-side copper tape (Agar Scientific, Stansted, UK). All XPS measurements were performed in constant analyzer energy (CAE) mode. Survey and high-resolution spectra (C1s, O1s, Si2p, Cu2p$_{3/2}$, N1s, Ca2p and Cl2p) were acquired at a pass energy of 150 and 100 eV, respectively, and with a step size of 1.0 and 0.1 eV, respectively. Detailed spectra processing was performed by commercial Thermo Avantage software (v. 4.75© 1999–2010 Thermo Fisher Scientific, Waltham, MA,

USA). Curve-fitting analysis was applied to the Cu2p$_{3/2}$ high-resolution spectra, in order to assess Cu chemical state. Surface atomic percentages were determined after Shirley background removal, using *Scofield* sensitivity factors. The same peak lineshape parameters (Gaussian/Lorentzian ratio and full width at half maximum) values were employed for the curve fitting of components belonging to the same high-resolution spectrum. Spectra were corrected for charge compensation effects by offsetting the binding energy relative to the aliphatic component of the C1s spectrum, which was set to 284.8 eV.

3.5. Kinetics of Copper Release in Aqueous Solution by Electro-Thermal Atomic Absorption Spectroscopy (ETAAS)

Copper release from CuNP-modified polyurethane foams was evaluated by electro-thermal atomic absorption spectroscopy (ETAAS). Cu quantification was achieved by means of a calibration curve obtained by the analysis, carried out in triplicate and in random sequence, of standard solutions at known Cu concentration, prepared by subsequent dilution of a stock Cu commercial solution (Fluka, Milan, Italy, Copper Standard for AAS TraceCERT®, 1000 mg/L Cu in nitric acid). Each PU sample was immersed in a Pyrex bottle containing 25 mL of contact solution. The latter was obtained as a 1:1 mixture of a 0.85% *w/w* NaCl aqueous solution (Fluka, purity ⩾99.5%) with phosphate buffer (PBS) K$_2$HPO$_4$/NaH$_2$PO$_4$ with known pH and ionic strength (6.4 and 0.1, respectively). At fixed times (10 min, 1 h, 2 h, 4 h, 7 h, 10 h and 24 h), 200 µL of this solution were taken from each vessel. These samples were appropriately diluted and acidified by 0.7% HNO$_3$ solution; if needed, additional dilution was applied in order to keep the measured Cu concentration within the linearity interval of the calibration curve. Samples were analyzed by a Perkin-Elmer dual-beam spectrophotometer, model 460 (Milan, Italy), using electro-thermal atomization in graphite furnace. Source was a hollow-cathode lamp (absorption line at 324.7 nm). The signal acquisition mode provided an automatic background correction by a deuterium lamp, in order to eliminate possible matrix absorption overlapping with the Cu absorption band. The entire kinetic experiment was carried out at 25 °C. The measured data were fitted by a first order kinetic Equation (1), where $[Cu]^0$ represented the amount of copper ions immediately dissolved (and therefore immediately available) in solution, $[Cu]^{max}$ was the maximum copper concentration in solution reached within 24 h, and k was the release kinetic constant. Data analysis was carried out using SigmaPlot® 12.0 software (Systat Software, San Jose, CA, USA).

$$[Cu] = [Cu]^0 + [Cu]^{max} \left(1 - e^{-kt}\right) \tag{1}$$

The possible Cu release from both vessels/glassware and the untreated polyurethane foams was also assayed; and the resulting Cu concentrations were all below the limit of quantification (LOQ).

3.6. Antimicrobial Tests

Escherichia coli ATCC 25922, *Staphylococcus aureus* FDA 209P (MSSA, with methicillin resistance 0.125 µg/mL), and *Kluyveromyces marxianus* CBS 608 were selected as target microorganisms. Aliquots of the different lyophilized microorganisms were reconstituted in 0.9% NaCl solution, added to 20 mL of sterile nutrient broth (Oxoid) and incubated for 24 h at the optimal temperature for microbial growth, which was 37 °C for *Staphylococcus aureus*, 42 °C for *Escherichia coli*, and 28 °C for *Kluyveromyces marxianus*. Each sample was then diluted 10^5 times by saline solution (pH = 6.4, $[Cl^-]$ = 0.15 M). 5 mL of each culture broth were left in contact with each type of CuNP treated foams, and incubated for 24 h, at the respective optimal growth temperatures. Untreated samples were used as controls. After the selected incubation times, 1 mL of culture broth was taken from each vessel and inoculated in a Petri dish containing nutrient agar. After 24 h, bacterial colony count was performed on each Petri dish, in order to assess the entity of bacterial growth inhibition exerted by treated polyurethane foams.

Materials **2016**, *9*, 544

4. Conclusions

Sacrificial-anode electrochemical synthesis of TOAC-stabilized CuNPs was used to modify industrial polyurethane foams. TEM and XPS analyses on Cu-nanocolloids demonstrated that the surfactant employed was capable of stabilizing freshly dispersed CuNPs from both the morphological and the chemical point of view. XPS measurements on copper-polyurethane composites revealed that a simple impregnation protocol employing diluted colloids was effective in modifying industrial polyurethane foams with CuNPs. These data, along with copper ion release measurements, showed that the final copper surface availability, along with the release of antibacterial ions in physiological solution, could be tuned just by changing the CuNP concentration in the impregnation baths. Foam pore size and aging resulted to affect the ion release to a minor extent, although samples stored for two months were demonstrated to be still very active in releasing bioactive copper ions. Biological tests showed that the proposed nano-functionalized materials exert a marked inhibitory effect on the growth of different target microorganisms such as *S. aureus*, *E. coli* and *K. marxianus*.

Acknowledgments: Lino Mondino, R & D director of Adler Group, is gratefully acknowledged for technical discussions and for kindly providing samples of polyurethane foams. Leonardo Bellomo and Vito Scannicchio are acknowledged for partial contribution to the collection of some experimental data. Financial support from Italian MIUR (Project "Silver" PON01_02210) and from Regione Puglia (Project code 56: "Laboratorio di tecnologie di modificazione superficiale di fibre naturali per il rilancio del settore tessile in Puglia") is gratefully acknowledged.

Author Contributions: M.C.S. contributed to perform most of the experiments, and wrote the first draft of the paper. R.A.P. designed the experiments and analyzed the data. R.R. performed release experiments and contributed to other experimental activities. E.B. and G.T. performed antimicrobial tests. M.P., A.S., and A.V. contributed to materials characterization and to critical discussions. T.R.I.C. and N.C. supervised research activities and defined the final version of the manuscript. N.C. coordinated the project on Cu-modified polyurethanes. Authorship is limited to those who have contributed substantially to the work reported.

Conflicts of Interest: The authors declare no conflict of interest. The founding sponsors had no role in the design of the study; in the collection, analyses, or interpretation of data; in the writing of the manuscript, and in the decision to publish the results.

Abbreviations

The following abbreviations are used in this manuscript:

BE	binding energy
CFU	colony forming unit
CNT	carbon nanotube
CuNP	copper nanoparticle
ETAAS	electro-thermal atomic absorption spectroscopy
PBS	phosphate buffer saline
PU	polyurethane
SAE	sacrificial anode electrolysis
TEM	transmission electron microscopy

References

1. Morent, R.; De Geyter, N. Improved textile functionality through surface modifications. In *Functional Textiles for Improved Performance, Protection and Health*; Pan, N., Sun, G., Eds.; Woodhead Publishing Series in Textiles; Woodhead Publishing: Cambridge, UK, 2011; pp. 3–26.
2. Gao, Y.; Cranston, R. Recent Advances in Antimicrobial Treatments of Textiles. *Text. Res. J.* **2008**, *78*, 60–72.
3. García, B.; Saiz-Poseu, J.; Gras-Charles, R.; Hernando, J.; Alibés, R.; Novio, F.; Sedó, J.; Busqué, F.; Ruiz-Molina, D. Mussel-Inspired Hydrophobic Coatings for Water-Repellent Textiles and Oil Removal. *ACS Appl. Mater. Interfaces* **2014**, *6*, 17616–17625. [CrossRef] [PubMed]
4. Onar, N.; Mete, G. Development of water-, oil-repellent and flame-retardant cotton fabrics by organic-inorganic hybrid materials. *J. Text. Inst.* **2016**. [CrossRef]

5.   Badanova, A.K.; Kutzhanova, A.Z.; Krichevsky, G.E. Research of the influence of hydrophobic finishing on coloristic characteristics of cellulosic textile material. *Izv. Vysshikh Uchebnykh Zaved. Seriya Teknol. Tekstil'noi Promyshlennosti* **2015**, *2015*, 63–66.

6.   Baltušnikaite, J.; Varnaite-Žuravliova, S.; Rubežiene, V.; Rimkute, R.; Verbiene, R. Influence of silver coated yarn distribution on electrical and shielding properties of flax woven fabrics. *Fibres Text. East. Eur.* **2014**, *104*, 84–90.

7.   Hu, C.-C.; Chang, S.-S.; Liang, N.-Y. Preparation and characterization of carbon black/polybutylene terephthalate/polyethylene terephthalate antistatic fiber with sheath–core structure. *J. Text. Inst.* **2015**. [CrossRef]

8.   Baseri, S. Preparation and characterization of conductive and antibacterial polyacrylonitrile terpolymer yarns produced by one-step organic coating. *J. Text. Inst.* **2016**. [CrossRef]

9.   Patra, J.K.; Gouda, S. Application of nanotechnology in textile engineering: An overview. *J. Eng. Technol. Res.* **2013**, *5*, 104–111. [CrossRef]

10.   Chandramohan, D.; Marimuthu, K. A review on natural fibers. *Int. J. Res. Rev. Appl. Sci.* **2011**, *8*, 194–206.

11.   Seves, A.; Romanò, M.; Maifreni, T.; Sora, S.; Ciferri, O. The microbial degradation of silk: A laboratory investigation. *Int. Biodeterior. Biodegrad.* **1998**, *42*, 203–211. [CrossRef]

12.   Szostak-Kotowa, J. Biodeterioration of textiles. *Int. Biodeterior. Biodegrad.* **2004**, *53*, 165–170. [CrossRef]

13.   Cioffi, N.; Rai, M. *Nano-Antimicrobials: Progress and Prospects*, 1st ed.; Springer: Berlin, Germany, 2012.

14.   Giannossa, L.C.; Longano, D.; Ditaranto, N.; Nitti, M.A.; Paladini, F.; Pollini, M.; Rai, M.; Sannino, A.; Valentini, A.; Cioffi, N. Metal nanoantimicrobials for textile applications. *Nanotechnol. Rev.* **2013**, *2*, 307–331. [CrossRef]

15.   Torres, A.; Ruales, C.; Pulgarin, C.; Aimable, A.; Bowen, P.; Sarria, V.; Kiwi, J. Innovative high-surface-area CuO pretreated cotton effective in bacterial inactivation under visible light. *ACS Appl. Mater. Interfaces* **2010**, *2*, 2547–2552. [CrossRef] [PubMed]

16.   Teli, M.D.; Sheikh, J. Bamboo rayon-copper nanoparticle composites as durable antibacterial textile materials. *Compos. Interfaces* **2014**, *21*, 161–171. [CrossRef]

17.   Subramanian, B.; Anu Priya, K.; Thanka Rajan, S.; Dhandapani, P.; Jayachandran, M. Antimicrobial activity of sputtered nanocrystalline CuO impregnated fabrics. *Mater. Lett.* **2014**, *128*, 1–4. [CrossRef]

18.   Perelshtein, I.; Applerot, G.; Perkas, N.; Wehrschuetz-Sigl, E.; Hasmann, A.; Guebitz, G.; Gedanken, A. CuO-cotton nanocomposite: Formation, morphology, and antibacterial activity. *Surf. Coat. Technol.* **2009**, *204*, 54–57. [CrossRef]

19.   Sedighi, A.; Montazer, M.; Samadi, N. Synthesis of nano Cu_2O on cotton: Morphological, physical, biological and optical sensing characterizations. *Carbohydr. Polym.* **2014**, *110*, 489–498. [CrossRef] [PubMed]

20.   Gouda, M.; Hebeish, A. Preparation and evaluation of CuO/Chitosan nanocomposite for antibacterial finishing cotton fabric. *J. Ind. Text.* **2010**, *39*, 203–214. [CrossRef]

21.   Beddow, J.; Singh, G.; Blanes, M.; Molla, K.; Perelshtein, I.; Gedanken, A.; Joyce, E.; Mason, T. Sonochemical coating of textile fabrics with antibacterial nanoparticles. In AIP Conference Proceedings; AIP Publishing: Melville, NY, USA, 2012; pp. 400–403.

22.   Anita, S.; Ramachandran, T.; Rajendran, R.; Koushik, C.V.; Mahalakshmi, M. A study of the antimicrobial property of encapsulated copper oxide nanoparticles on cotton fabric. *Text. Res. J.* **2011**, *81*, 1081–1088. [CrossRef]

23.   Abramov, O.V.; Gedanken, A.; Koltypin, Y.; Perkas, N.; Perelshtein, I.; Joyce, E.; Mason, T.J. Pilot scale sonochemical coating of nanoparticles onto textiles to produce biocidal fabrics. *Surf. Coat. Technol.* **2009**, *204*, 718–722. [CrossRef]

24.   Castro, C.; Sanjines, R.; Pulgarin, C.; Osorio, P.; Giraldo, S.A.; Kiwi, J. Structure-reactivity relations for DC-magnetron sputtered Cu-layers during *E. coli* inactivation in the dark and under light. *J. Photochem. Photobiol. A Chem.* **2010**, *216*, 295–302. [CrossRef]

25.   Sedighi, A.; Montazer, M.; Hemmatinejad, N. Copper nanoparticles on bleached cotton fabric: In situ synthesis and characterization. *Cellulose* **2014**, *21*, 2119–2132. [CrossRef]

26.   Gabbay, J. Copper oxide impregnated textiles with potent biocidal activities. *J. Ind. Text.* **2006**, *35*, 323–335. [CrossRef]

27.   Borkow, G.; Gabbay, J. Putting copper into action: Copper-impregnated products with potent biocidal activities. *FASEB J.* **2004**, *18*, 1728–1730. [CrossRef] [PubMed]

28. Borkow, G.; Gabbay, J. Copper, an ancient remedy returning to fight microbial, fungal and viral infections. *Curr. Chem. Biol.* **2009**, *3*, 272–278. [CrossRef]

29. Huang, Z.-M.; Zhang, Y.-Z.; Kotaki, M.; Ramakrishna, S. A review on polymer nanofibers by electrospinning and their applications in nanocomposites. *Compos. Sci. Technol.* **2003**, *63*, 2223–2253. [CrossRef]

30. Brzeziński, S.; Malinowska, G.; Kowalczyk, D.; Kaleta, A.; Boak, B.; Jasiorski, M.; Dąbek, K.; Baszczuk, A.; Tracz, A. Antibacterial and fungicidal coating of textile-polymeric materials filled with bioactive nano- and submicro-particles. *Fibers Text. East. Eur.* **2012**, *20*, 70–77.

31. Mumcuoglu, K.Y.; Gabbay, J.; Borkow, G. Copper oxide-impregnated fabrics for the control of house dust mites. *Int. J. Pest Manag.* **2008**, *54*, 235–240. [CrossRef]

32. Rio, L.; Kusiak-Nejman, E.; Kiwi, J.; Bétrisey, B.; Pulgarin, C.; Trampuz, A.; Bizzini, A. Comparison of methods for evaluation of the bactericidal activity of copper-sputtered surfaces against methicillin-resistant *Staphylococcus aureus*. *Appl. Environ. Microbiol.* **2012**, *78*, 8176–8182. [CrossRef] [PubMed]

33. Komeily-Nia, Z.; Montazer, M.; Latifi, M. Synthesis of nano copper/nylon composite using ascorbic acid and CTAB. *Colloids Surf. A Physicochem. Eng. Asp.* **2013**, *439*, 167–175. [CrossRef]

34. Amina, M.; Amna, T.; Hassan, M.S.; Ibrahim, T.A.; Khil, M.-S. Facile single mode electrospinning way for fabrication of natural product based silver decorated polyurethane nanofibrous membranes: Prospective medicated bandages. *Colloids Surf. A Physicochem. Eng. Asp.* **2013**, *425*, 115–121. [CrossRef]

35. Tijing, L.D.; Ruelo, M.T.G.; Amarjargal, A.; Pant, H.R.; Park, C.-H.; Kim, C.S. One-step fabrication of antibacterial (silver nanoparticles/poly(ethylene oxide))—Polyurethane bicomponent hybrid nanofibrous mat by dual-spinneret electrospinning. *Mater. Chem. Phys.* **2012**, *134*, 557–561. [CrossRef]

36. Nirmala, R.; Kalpana, D.; Navamathavan, R.; Park, M.; Kim, H.Y.; Park, S.-J. Antimicrobial activity of electrospun polyurethane nanofibers containing composite materials. *Korean J. Chem. Eng.* **2014**, *31*, 855–860. [CrossRef]

37. Jeon, H.J.; Kim, J.S.; Kim, T.G.; Kim, J.H.; Yu, W.-R.; Youk, J.H. Preparation of poly(ε-caprolactone)-based polyurethane nanofibers containing silver nanoparticles. *Appl. Surf. Sci.* **2008**, *254*, 5886–5890. [CrossRef]

38. Prabhakar, P.K.; Raj, S.; Anuradha, P.R.; Sawant, S.N.; Doble, M. Biocompatibility studies on polyaniline and polyaniline-silver nanoparticle coated polyurethane composite. *Colloids Surf. B Biointerfaces* **2011**, *86*, 146–153. [CrossRef] [PubMed]

39. Yang, Z.; Qiu, S.; Wang, Y.; Lv, H.; Xing, X.; Luo, J. Synthesis and characterization of re-dispersible silver nanoparticles/polyurethane hybrid materials. *Polym. Mater. Sci. Eng.* **2012**, *28*, 118–121.

40. Toker, R.D.; Kayaman-Apohan, N.; Kahraman, M.V. UV-curable nano-silver containing polyurethane based organic-inorganic hybrid coatings. *Prog. Org. Coat.* **2013**, *76*, 1243–1250. [CrossRef]

41. Pant, H.R.; Kim, H.J.; Joshi, M.K.; Pant, B.; Park, C.H.; Kim, J.I.; Hui, K.S.; Kim, C.S. One-step fabrication of multifunctional composite polyurethane spider-web-like nanofibrous membrane for water purification. *J. Hazard. Mater.* **2014**, *264*, 25–33. [CrossRef] [PubMed]

42. Kim, J.H.; Unnithan, A.R.; Kim, H.J.; Tiwari, A.P.; Park, C.H.; Kim, C.S. Electrospun badger (Meles meles) oil/Ag nanoparticle based anti-bacterial mats for biomedical applications. *J. Ind. Eng. Chem.* **2015**, *30*, 254–260. [CrossRef]

43. Dumitriu, R.P.; Sacarescu, L.; Macocinschi, D.; Filip, D.; Vasile, C. Effect of silver nanoparticles on the dispersion, rheological properties and morphological aspect of solvent cast polyurethane/biopolymers bionanocomposite membranes. *J. Adhes. Sci. Technol.* **2016**, *30*, 1716–1726. [CrossRef]

44. Wang, X.; Chen, M.-Q.; Chen, Q.-H.; Lu, J.; Cheng, L.; Jiang, H.; Dai, L.-F.; Zhang, H.-D.; Yang, T.-W.; Pei, Y.-H.; et al. Reduction of biofilm formation in rabbits by novel nano-silver/polyurethane coated endotracheal tube. *J. Biomater. Tissue Eng.* **2015**, *5*, 961–966. [CrossRef]

45. Subagia, I.D.G.A.; Jiang, Z.; Tijing, L.D.; Kim, Y.; Kim, C.S.; Lim, J.K.; Lim, J.K. Hybrid multi-scale basalt fiber-epoxy composite laminate reinforced with Electrospun polyurethane nanofibers containing carbon nanotubes. *Fibers Polym.* **2014**, *15*, 1295–1302. [CrossRef]

46. Shamshi Hassan, M.; Amna, T.; Sheikh, F.A.; Al-Deyab, S.S.; Eun Choi, K.; Hwang, I.H.; Khil, M.-S. Bimetallic Zn/Ag doped polyurethane spider net composite nanofibers: A novel multipurpose electrospun mat. *Ceram. Int.* **2013**, *39*, 2503–2510. [CrossRef]

47. Tijing, L.D.; Ruelo, M.T.G.; Amarjargal, A.; Pant, H.R.; Park, C.-H.; Kim, D.W.; Kim, C.S. Antibacterial and superhydrophilic electrospun polyurethane nanocomposite fibers containing tourmaline nanoparticles. *Chem. Eng. J.* **2012**, *197*, 41–48. [CrossRef]

48. Tijing, L.D.; Amarjargal, A.; Jiang, Z.; Ruelo, M.T.G.; Park, C.-H.; Pant, H.R.; Kim, D.-W.; Lee, D.H.; Kim, C.S. Antibacterial tourmaline nanoparticles/polyurethane hybrid mat decorated with silver nanoparticles prepared by electrospinning and UV photoreduction. *Curr. Appl. Phys.* **2013**, *13*, 205–210. [CrossRef]

49. Luo, Z.; Hong, R.Y.; Xie, H.D.; Feng, W.G. One-step synthesis of functional silica nanoparticles for reinforcement of polyurethane coatings. *Powder Technol.* **2012**, *218*, 23–30. [CrossRef]

50. Amna, T.; Hassan, M.S.; Sheikh, F.A.; Lee, H.K.; Seo, K.-S.; Yoon, D.; Hwang, I.H. Zinc oxide-doped poly(urethane) spider web nanofibrous scaffold via one-step electrospinning: A novel matrix for tissue engineering. *Appl. Microbiol. Biotechnol.* **2013**, *97*, 1725–1734. [CrossRef] [PubMed]

51. Sheikh, F.A.; Kanjwal, M.A.; Saran, S.; Chung, W.-J.; Kim, H. Polyurethane nanofibers containing copper nanoparticles as future materials. *Appl. Surf. Sci.* **2011**, *257*, 3020–3026. [CrossRef]

52. Tian, Q.; Guo, X. Electroless copper plating on microcellular polyurethane foam. *Trans. Nonferrous Met. Soc. China* **2010**, *20*, s283–s287. [CrossRef]

53. Nirmala, R.; Jeon, K.S.; Lim, B.H.; Navamathavan, R.; Kim, H.Y. Preparation and characterization of copper oxide particles incorporated polyurethane composite nanofibers by electrospinning. *Ceram. Int.* **2013**, *39*, 9651–9658. [CrossRef]

54. Reetz, M.T.; Helbig, W. Size-selective synthesis of nanostructured transition metal clusters. *J. Am. Chem. Soc.* **1994**, *116*, 7401–7402. [CrossRef]

55. Cioffi, N.; Torsi, L.; Sabbatini, L.; Zambonin, P.G.; Bleve-Zacheo, T. Electrosynthesis and characterisation of nanostructured palladium-polypyrrole composites. *J. Electroanal. Chem.* **2000**, *488*, 42–47. [CrossRef]

56. Ditaranto, N.; Picca, R.A.; Sportelli, M.C.; Sabbatini, L.; Cioffi, N. Surface characterization of manufactured goods modified by metal/metal oxides nano-antimicrobials. *Surf. Interface Anal.* **2016**, *48*, 505–508. [CrossRef]

57. Wagner, C.D. *Handbook of X-ray Photoelectron Spectroscopy: A Reference Book of Standard Data for Use in X-ray Photoelectron Spectroscopy*; Physical Electronics Division; Perkin-Elmer Corp.: Waltham, MA, USA, 1979.

58. NIST XPS Database. Available online: http://www.srdata.nist.gov/xps (accessed on 5 July 2016).

59. Jirka, I. An ESCA study of copper clusters on carbon. *Surf. Sci.* **1990**, *232*, 307–315. [CrossRef]

60. Wu, Y.; Garfunkel, E.; Madey, T.E. Initial stages of Cu growth on ordered Al_2O_3 ultrathin films. *J. Vac. Sci. Technol. A* **1996**, *14*, 1662–1667. [CrossRef]

61. Cioffi, N.; Torsi, L.; Ditaranto, N.; Tantillo, G.; Ghibelli, L.; Sabbatini, L.; Bleve-Zacheo, T.; D'Alessio, M.; Zambonin, P.G.; Traversa, E. Copper Nanoparticle/Polymer Composites with Antifungal and Bacteriostatic Properties. *Chem. Mater.* **2005**, *17*, 5255–5262. [CrossRef]

62. Cioffi, N.; Torsi, L.; Ditaranto, N.; Sabbatini, L.; Zambonin, P.G.; Tantillo, G.; Ghibelli, L.; D'Alessio, M.; Bleve-Zacheo, T.; Traversa, E. Antifungal activity of polymer-based copper nanocomposite coatings. *Appl. Phys. Lett.* **2004**, *85*, 2417–2419. [CrossRef]

63. Cioffi, N.; Ditaranto, N.; Torsi, L.; Picca, R.A.; Giglio, E.D.; Sabbatini, L.; Novello, L.; Tantillo, G.; Bleve-Zacheo, T.; Zambonin, P.G. Synthesis, analytical characterization and bioactivity of Ag and Cu nanoparticles embedded in poly-vinyl-methyl-ketone films. *Anal. Bioanal. Chem.* **2005**, *382*, 1912–1918. [CrossRef] [PubMed]

64. ImageJ. Available online: http://imagej.nih.gov/ij/ (accessed on 19 November 2015).

materials

MDPI

Article

Exopolysaccharide-Based Bioflocculant Matrix of *Azotobacter chroococcum* XU1 for Synthesis of AgCl Nanoparticles and Its Application as a Novel Biocidal Nanobiomaterial

Bakhtiyor A. Rasulov [1], Parhat Rozi [2,3], Mohichehra A. Pattaeva [4], Abulimiti Yili [2] and Haji Akber Aisa [2,*]

[1]   Institute of Genetics and Plant Experimental Biology, Uzbekistan Academy of Sciences, Yukori Yuz, Kybray District 111226, Uzbekistan; bakhtiyor_1980@mail.ru

[2]   Key Laboratory of Plant Resources and Chemistry in Arid Region, Xinjiang Technical Institute of Chemistry and Physics, Chinese Academy of Sciences, Urumqi 830011, China; parhatruzi@126.com (P.R.); abu@ms.xjb.ac.cn (A.Y.)

[3]   University of Chinese Academy of Sciences, Beijing 100039, China

[4]   Institute of Microbiology, Uzbekistan Academy of Sciences, Tashkent 100128, Uzbekistan; maqsudaziz@mail.ru

*   Correspondence: haji@ms.xjb.ac.cn; Tel.: +86-138-9989-2388

Academic Editor: Mauro Pollini

Received: 24 March 2016; Accepted: 25 June 2016; Published: 29 June 2016

Abstract: A simple and green method was developed for the biosynthesis of AgCl nanoparticles, free from Ag nanoparticles, using the exopolysaccharide-based bioflocculant of nitrogen fixing *Azotobacter chroococcum* XU1 strain. AgCl nanoparticles were characterized by UV-Vis, X-ray diffraction (XRD), Fourier Transform-Infra Red (FT-IR) and Scanning electron microscopy-energy dispersive X-ray (SEM-EDX). The concentration-dependent and controllable method for the synthesis of AgCl nanoparticles of a certain size and morphology was developed. As-synthesized AgCl nanoparticles were characterized bya high content of AgCl and exhibited strong antimicrobial activity towards pathogenic microorganisms such as *E. coli*, *S. aureus* and *C. albicans*. The biofabricated AgCl nanoparticles can be exploited as a promising new biocidalbionanocomposite against pathogenic microorganisms.

Keywords: exopolysaccharide-based bioflocculant; AgCl nanoparticles; *Azotobacter chroococcum*; antimicrobial activity

1. Introduction

Bioflocculants are organic macromolecular substances secreted by a wide variety of microorganisms [1–6], and microbial bioflocculants have attracted considerable attention in recent years due to their biodegradable, harmless and negligible secondary pollution [1]. The polysaccharide-based bioflocculants are exploitedfor theirhigh efficiency in disposing of different of chemicals [7–10]. The polysaccharide macromolecules in bioflocculants contain monomeric units of sugar molecules such as glucose, mannose, fructose, or rhamnose, etc. [11]. The hydroxyl groups of these sugars strongly associate with metal ions leading to the formation of different types of biomaterials. Besides, they allow the control of shape, size and particle dispersion.

Polysaccharide-based bioflocculants easily form a variety of liquid crystals in aqueous solutions and bioflocculant-mediated processes are highly profitable. Polysaccharide-based bioflocculants been reported for the synthesis of silver nanoparticles (AgNPs). Along with AgNPs, silver chloride nanoparticles (AgCl-NPs) have alsoreceived wide interest. Electrochemical, chemical reduction, surfactants, photochemical reduction, ionic liquid microemulsion, ultrasound irradiation, hydrolysis and ion-exchange reactions and thermal decomposition methods were developed for the synthesis of AgNPs and AgCl-NPs [12–16]. In recent years, the synthesis of silver halides (AgX) received considerable attention since they are highly photosensitive semiconductors. They have been extensively used as photosensitizers andsource materials in photographic films and can also be used as metallic silver precursors [17–19]. A total "biological" method for the synthesis of face-centered cubic Ag/AgCl-NPs was documented, using the cellulose of a non-pathogenic *Gluconacetobacter xylinum*. As-synthesized nanocomposite exhibited considerable antibacterial activity against *S. aureus* and *E. coli* [20]. Besides, the chitosan oligomers [21], and bacterial [22] and non-bacterial cellulose were also used as a template to synthesize Ag/AgCl-NPs [23]. These nanocomposites also displayed high antibacterial activity [22]. Moreover, a controlled synthesis of AgCl-NPs by the polysaccharide-based bioflocculant of diazotrophic strain *Bradyrhizobium japonicum* 36 was reported recently, where at equal molar concentrations of Ag^+ and Cl^-, the polysaccharide-based bioflocculants act as a template to assemble only AgCl-NPs with an average diameter of 10 nm [24].

In this study we report the green and controlled synthesis method of AgCl-NPs using the expolysaccharide-based bioflocculant of diazotrophic nitrogen-fixing bacterial strain *Azotobacter chroococcum* XU1. As-synthesized AgCl-NPs further tested as a biocidal agent against pathogenic microorganisms, namely *Candida albicans*, *Escherichia coli* and *Staphylococcus aureus*.

2. Results

2.1. Biosynthesis and Characterization of AgCl-NPs

Asit was documented, AgCl-NPs develop from a milk-white to a yellowish-brown color in aqueous solution due to excitation of surface plasmon resonance (SPR) [25,26]. Theirsynthesis can be monitored by UV-visible absorption spectra. AgCl-NPs exhibit a broad and strong absorption at 200–350 nm, which can be ascribed to the characteristic absorption of the AgCl semiconductors [27]. The shift in the band may have contributed to the different sizes of the AgCl nanoparticles. Since bioflocculants basically are composed of polysaccharide macromolecules, it is highly possible to reduce the silver ions and stabilize them into the polysaccharide matrix. This evidence makes the parallel synthesis of both AgCl-NPs and AgNPs in the presence of chloride ions and polysaccharide matrix inevitable [21–23]. To biofabricate only AgCl-NPs, a concentration-dependent method was evaluated. It was observed that the intensive synthesis of AgCl occurred in low concentrations of Ag^+ in aqueous solution. The polysaccharide matrix in the reaction mixture enabled the formation of AgCl-NPs with absorption bands at 250–260 nm (Figure 1). This resulted in the absence of absorption within 380–550 nm in the visible-light region, which indicates no synthesis of AgNPs. At a high concentration of Ag^+, as shown in Figure 1, the formation of both AgCl-NPs and AgNPs was observed (Figure 1; absorption band at 260 and 435 nm assigned as a purple chart). The appearance of the surface plasmon resonance (SPR) transition at 435 nm confirms the production of AgNPs [28,29].

It can be concluded that synthesis of AgCl-NPs and AgNPs is concentration-dependent and the formation of nanoparticles is controllable. For the synthesis of only AgCl-NPs, the concentration of chloride ions plays a key role. Their concentration must be always higher than that of Ag ions in the reaction mixture. However, it is necessary to point out that the AgCl nanoparticles are aggregated into large clusters, which prevented an accurate estimation of their morphology and size distributions [30]. An analogical conclusion was made in previous reports onthe controlled synthesis of AgCl-NPs using the bioflocculant of *Bradyrhizobium japonicum* [24]. Authors reported that when the Ag^+ concentration exceeded that of Cl^-, the formation of AgCl-NPs/AgNPs was observed, and in the presence of Cl^- ions, the synthesis of AgCl and its assembly over the polysaccharide-based bioflocculant occurs first. When the chloride ions were depleted in the reaction mixture, the excess amount of silver ions was reduced and stabilized in the polysaccharide-based bioflocculant, leading the formation of AgCl-NPs/AgNPs [24].

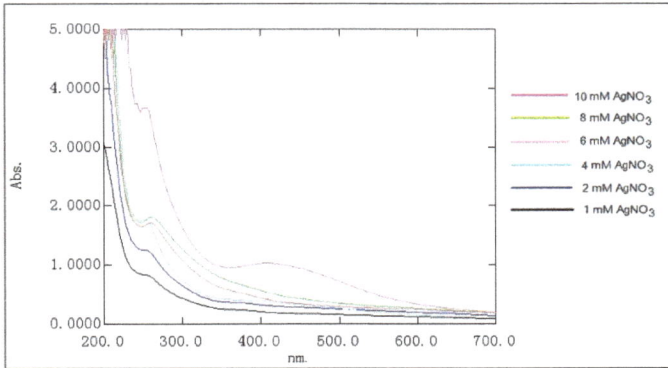

Figure 1. UV-vis spectra of AgCl-NPs in different concentration of AgNO$_3$.

2.2. SEM-EDX

The SEM micrographs of AgCl-NPs clearly showed rough agglomerations of polydisperse character and ranges of approximately 10 to 68 nm (Figure 2).

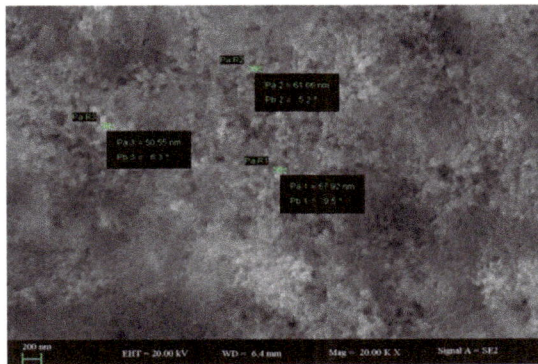

(a)

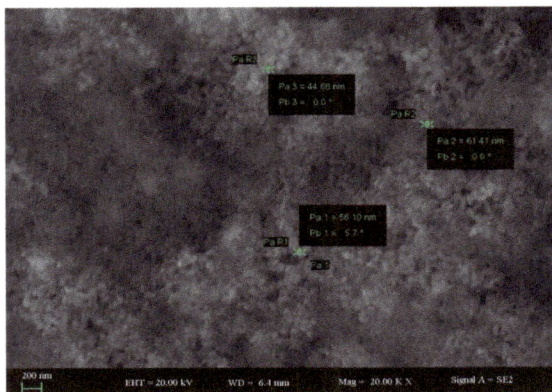

(b)

Figure 2. (a,b) SEM images of AgCl-NPs.

The Energy dispersive X-ray (EDX) spectra were recorded in order to provide further confirmation on the formation of AgCl-NPs on exopolysaccharide template. The *x*- and *y*-axis labels were energy/keV and count/cps, respectively [30]. The EDX spectra of AgCl-NPs–loaded exopolysaccharide nanocomposite confirm the existence of silver and chlorine atoms. As was seen in the EDX profile, EPS-stabilized AgCl-NPs showed strong signals at ~3 keV for silver atoms, and at ~2.5 keV for chlorine atoms (Figure 3). A semi-quantitative analysis showed the atomic ratio of the Ag and Cl elements was approximately 1:1. This evidence confirms the theoretic stoichiometric atomic ratio between Ag and Cl atoms in AgCl. The data were in agreement with those documented elsewhere [30].

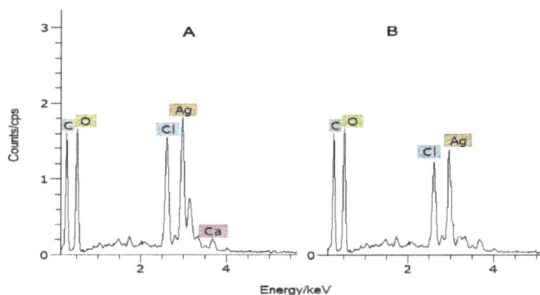

Figure 3. EDX spectra of AgCl-NPs, synthesized using Ag^+ and Cl^- in different concentrations of Ag^+ ((**A**)—8 mM $AgNO_3$; (**B**)—6 mM $AgNO_3$).

2.3. X-ray Diffraction Analysis of AgCl-NPs

AgCl-NPs were characterized using the X-ray diffractometer with Cu-Kα radiation operated at a voltage of 40 kV in the scan range of 20 to 80 degree. The XRD pattern of AgCl-NPs exhibited several size-dependent features leading to peak position, height and width. The XRD patterns of AgCl-NPs nanostructures are presented in Figure 4. The distinct diffraction peaks at a 2θ of 27.64°, 32.24°, 46.2°, 54.78°, 57.44°, 67.42°, 74.4° and 76.6° can be attributed to the (111), (200), (220), (311), (222), (400), (331) and (420) planes, which are the typical cubic phase of AgCl crystal (JCPDS file 31-1238) [27,30]. The strong and narrow diffraction peaks in some samples (obtained from relatively high concentrations of Ag^+ solutions) reveal the highly crystalline structure [27].

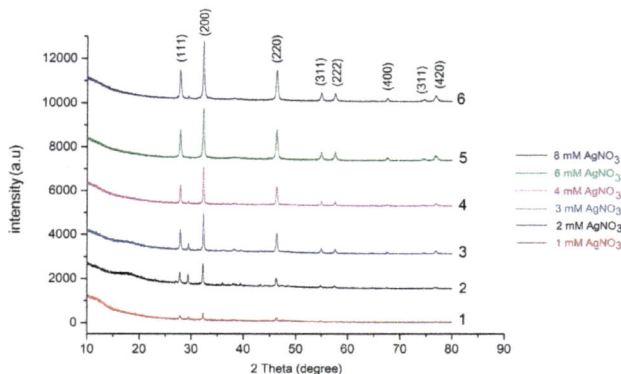

Figure 4. X-ray diffraction (XRD) pattern of AgCl-NPs, synthesized in different concentrations of $AgNO_3$.

Analogical data previously reported by other authors [27,30] suggest the existence of AgCl species in the synthesized nanostructures with the above-mentioned data. Revealed diffraction peaks were sharp and intense, indicating a high degree of crystallinity of the AgCl species.

2.4. FT-IR of AgCl-NPs

FT-IR estimation was carried out to investigate the possible functional groups and chemical bonds of AgCl-NPs containing nanobiocomposite. The FT-IR spectra of AgCl-NPs exhibited various characteristic peaks with ranges from 3420 to 800 cm^{-1} (Figure 5). The broadest and strongest peak was observed at 3420 cm^{-1}, due to a stretching vibration of the hydroxyl groups (O–H). The broad peaks at 1236–1063 cm^{-1} correspond to the C–O–C stretching from the glycosidic linkages and the O–H bending from alcohols [31,32], whereas the intense peak at 995 cm^{-1} confirms the presence of carbohydrates. Furthermore, weak absorption peaks at 820–972 cm^{-1} were observed, which confirmed that linkages had occurred between monosaccharides in a polysaccharide matrix of the bioflocculant.

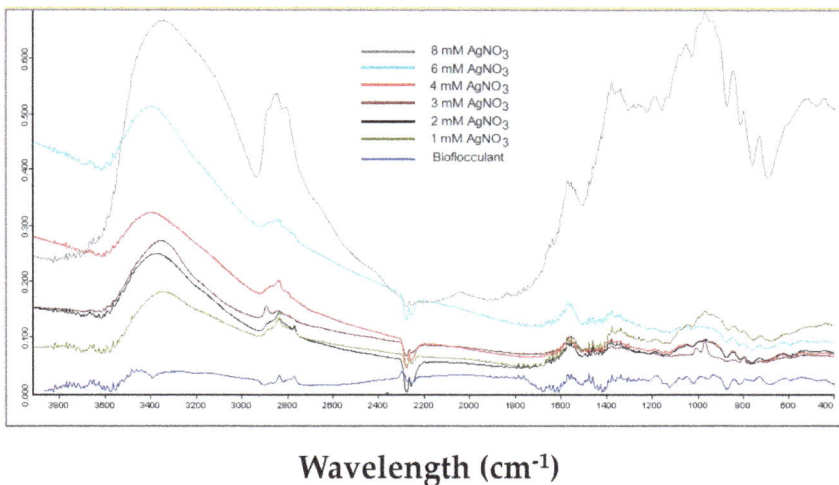

Wavelength (cm^{-1})

Figure 5. FT-IR spectrum of AgCl-NPs, assembled on the polysaccharide-based bioflocculant of *A. chroococcum* XU1 in different concentrations of Ag$^+$.

In some cases, when the polysaccharide matrix is used for the impregnation of nanoparticles, the absorption maxima of polysaccharides remain unaffected [32,33]. However, when the polysaccharide template is used as a reducing and stabilizing agent, slight changes in the absorption maxima of polysaccharides are observed. Kanmani and Lim (2013) report additional absorption peaks of bacterial exopolysaccharides at 2352 and 1449 cm^{-1}, which were attributed to the presence of the AgNPs [26]. The biosynthesized AgNPs' influence on the O–H stretching at 3428 cm^{-1} was also reported [1]. This evidence indicates that Ag/AgCl-NPs' association with biomolecules is varied due to the nature of the matrix and chemical bonds.

2.5. Antimicrobial Activity of AgCl-NPs (the Hole Method)

The antibacterial and antifungal activity of AgCl-NPs, fabricated using the polysaccharide-based bioflocculant of *A. chroococcum* XU1, was investigated against the pathogen bacterial strains *E. coli* ATCC11229, *S. aureus* ATCC6538 and pathogenic fungi *Candida albicans* ATCC1023. The AgCl-NPs samples were tested using the agar well diffusion method and via measuring the radial diameter of inhibition zones. Different concentrations of AgCl-NPs were applied, from 0.1 to 5 mg/mL. The results revealed that all pathogens were inhibited after treatment with AgCl-NPs. The inhibition rate was proportionally dependent on not only the concentration (dilution) of AgCl-NPs, but also on the concentration of the bioflocculant and AgNO$_3$, which were used to produce AgCl-NPs. The increase in the AgCl-NPs' antibacterial activity towards *E. coli* ATCC11229 and *S. aureus* ATCC6538 is shown in Table 1.

Table 1. Antibacterial activity of AgCl-NPs in 2.5 mg/mL concentration.

Strains	Inhibition Zone, mm
E. coli ATCC11229	22
S. aureus ATCC6538	23
C. albicans ATCC10231	15

The maximum antibacterial activity was observed with 2.5 mg/mL AgCl-NPs and it was several times higher than that of 1 mM of AgNO$_3$ applied as a control. Obtained data are in good agreement with those reports, where the dose-dependent impact of Ag/AgCl-NPs was evaluated [20–23]. The antifungal activity of nanoparticles also widely discussed in the literature and the impact of nanoparticles was evaluated against some genus of fungi, such as *Candida*, *Aspergillus* and *Penicillium* [26]. Some authors reported treatment with low concentrations of AgNPs. Kora and Arunachalam (2011) report the effective suppression of 4 µg /mL AgNPs against *P. aeruginosa* [34], whereas Wei et al. (2012) report 9 µg/mL AgNPs against *B. subtilis* and *E. coli* [35]. Bankura et al. (2012) tested higher concentrations of AgNPs [28]. At a 200 µg/mL concentration of AgNPs, the inhibition zone diameters of *E. coli*, *P. aeruginosa*, *B. cereus*, and *B. subtilis* were 21, 24, 28, and 32 mm, respectively. Similar results were reported by Panacek et al. (2008) [36]. Analogical results were obtained in our previous research, when the AgCl-NP concentrations ranged from 0.25 to 2.5 µg/mL. In these concentrations, the maximal inhibition zones in *C. albicans* ATCC10231 were from 7.5 to 15 mm [24]. The maximum antibacterial activity was observed with 2.5 mg/mL AgCl-NPs and it was 2.5–2.7 times higher than 1 mM AgNO$_3$ applied as a control. A similar dose-dependent impact of Ag/AgCl-NPs was also evaluated against *E. coli* and *S. aureus* [20,23].

In our research, the pathogenic strain *C. albicans* ATCC10231 was inhibited by AgCl-NPs, and the antifungal activity proportionally correlated with the concentration of AgCl-NPs applied. With 0.1 mg/mL of AgCl-NPs, the inhibition zone was 7.5 mm; when concentration reached 2.5 mg/mL, the inhibition zone was 15 mm.

3. Discussion

Several methods and mechanisms for the green synthesis of AgCl-NPs have been reported. Apart from the application of polysaccharides [20–23], other reducing agents are also used to reduce Ag$^+$ and to synthesize AgCl-NPs. Duran et al. (2014) [37] reported the biogenic synthesis of AgCl-NPs with laccase, whereas Awwad et al. (2015) applied the flowers of *Albizia julibrissin* extract, rich in chlorine ions, as a reducing, chlorinated and capping agent in the formation of Ag/AgCl-NPs [38].

In our research, the exopolysaccharide-based bioflocculant of *A. chroococcum* XU1 was used for the synthesis and stabilization of AgCl-NPs. The exopolysaccharide-based bioflocculant can reduce Ag$^+$ from AgNO$_3$, and synthesize AgCl-NPs and AgNPs with separate absorption maximums at 250–260 and 435 nm. It was found that the formation of AgCl-NPs and AgNPs was concentration-dependent. In earlier reports, the formation of a mix of AgCl-NPs/AgNPs was documented, since in most cases the concentration-dependent synthesis had not been evaluated [18,19,23,27]. In some reports, chlorine ions were randomly chosen, or controlling theconcentration was not an aim of the surveys [38]. In this report we tried to strictly control the formation of only one type of nanoparticle—AgCl-NP—and the synthesis was carried out atcertain concentrations of Ag$^+$ and Cl$^-$. At a fixed concentration of the exopolysaccharide-based bioflocculant (11.5 mg/mL) with 1 to 8 mM of Ag$^+$ (at the same concentrations of Cl$^-$), the enhanced synthesis of AgCl-NPs was observed. When the Ag$^+$ concentration exceeded that of Cl^-, the formation of AgCl-NPs/AgNPs was observed. It is supposed that in the presence of Cl^- ions, the synthesis of AgCl and its assembly over the exopolysaccharide-based bioflocculant occurs first. When the chloride ions were depleted in the reaction mixture, an excess amount of silver ions was reduced and stabilized in the exopolysaccharide-based bioflocculant, leading to the formation of AgCl-NPs/AgNPs. Concentration dependent and controlled synthesis of AgCl-NPs was also reported in our previous study [24]. The polysaccharide-based bioflocculant of *B. japonicum* 36 can reduce Ag^+ from $AgNO_3$, and synthesize AgCl-NPs and AgNPs with separate absorption maximums at 335 and 420 nm. It was found that the formation of AgCl-NPs and AgNPs was concentration-dependent. From 1 to 7 mM Ag^+, the formation of only AgCl-NPs was observed and it exhibited an absorption band at 335 nm. Further increasing the Ag^+ concentration led to the formation of both AgCl-NPs (335 nm) and AgNPs (420 nm). When the concentrations of both Ag^+ and Cl^- were the same, the AgCl-NPs were the only product, with an absorption band at 335 nm [24].

The XRD analysis of AgCl-NPs exhibited characteristic peaks at $27.64°$, $32.24°$, $46.2°$, $54.78°$, $57.44°$, $67.42°$, $74.4°$ and $76.6°$ which were attributed to planes (111), (200), (220), (311), (222), (400), (331), and (420) of the cubic phase of the AgCl crystal, and are in good agreement with early reports [27,31]. Further SEM-EDX and FT-IR analyses also confirmed the presence of nanoparticles and their interaction with some functional groups. The FT-IR analysis showed some new evidence on the influence of AgCl-NPs on some functional groups. Obtained data varies with those from earlier reports with AgNPs [1,24,26,33,36].

As-synthesized AgCl-NPs were further examined for antimicrobial activity towards *E. coli* ATCC11229, *S. aureus* ATCC6538 and *Candida albicans* ATCC1023, and their high biocidal activity was observed. The antibacterial and antifungal activity of AgCl-NPs was also widely reported and discussed in recent surveys [20–24]. Despite the AgCl-NPs being synthesized via different reduction methods and exploiting a wide range of templates, they exhibited considerable antimicrobial activity toward pathogens, such *E. coli*, *S. aureus* and *C. albicans* [20–24]. It can be concluded from the results that the AgCl nanoparticles biofabricated on the basis of the exopolysaccharide-based bioflocculant of *Azotobacter chroococcum* XU1 can be exploited as a promising new biocidal bionanocomposite against pathogenic microorganisms.

4. Materials and Methods

4.1. Materials and Microbial Strains

The polysaccharide-based bioflocculant producing diazotrophic strain *A. chroococcum* XU1, and pathogenic strains *Candida albicans* ATCC10231, *Escherichia coli* ATCC11229, *Staphylococcus aureus* ATCC6538 were obtained from the Culture Collection of State Key Laboratory Basis of Xinjiang Indigenous Medicinal Plants Resource Utilization, Xinjiang Technical Institute of Physics and Chemistry, CAS.

All solutions were made of using ultra filtered high purity deionized water. Reagents and chemicals used in this study were of analytical grade.

4.2. Production of Bioflocculant

Production and purification of bioflocculant by *A. chroococcum* XU1 was developed in our previous experiments [24,25,39] and its fractionation into protein, low- and high-molar-weight polysaccharides was used as a part of protocol for green synthesis of AgNPs [24,25]. Production of bioflocculant carried out in 2 L flasks containing 1 L of the modified medium with 150 rpm at 30 °C under intensive aeration. The composition of the modified medium was as follows: sucrose, 20 g/L; $MgSO_4 \cdot 7H_2O$, 0.2 g/L; KH_2PO_4, 0.2 g/L; NaCl, 0.1 g/L; $CaCO_3$, 10 g/L. The initial pH value of the medium was adjusted to 7.0. Each flask was inoculated with 4% (*v/v*) of the seed culture and incubated at 30 °C with shaking at 150 rpm for three days. Samples were withdrawn at different time intervals and monitored for cell growth and flocculating activity. Culture broth was centrifuged at 10,000 rpm for 15 min to separate the cells which were washed twice with distilled water.

4.3. Purification of Bioflocculant

Bioflocculant purification was carried out as documented elsewhere [1]. The culture broth was centrifuged at 10,000 rpm for 20 min. To the supernatant were added three volumes of cold ethanol instantly, until white cotton-like flocks were formed. The precipitate centrifuged at 10,000 rpm for 15 min. Then, the bioflocculant dialyzed against de-ionized water at 4 °C overnight to obtain purified bioflocculant, free from minerals.

4.4. Synthesis of AgCl-NPs

Preparation of AgCl-NPs was carried by a method, developed by us [24]. Freshly prepared bioflocculant was mixed with 10 mL of 2 mM NaCl solution with a vaccination needle and exposed to 10 mM $AgNO_3$ solution. The reaction mixture was left overnight at room temperature. For FT-IR and XRD analysis the reaction solution was centrifuged at 15,000 rpm for 5 min to obtain AgCl-NPs. The prepared AgCl-NPs were washed four times with distilled water to remove the impurities absorbed on the surface of AgCl-NPs.

4.5. Characterization of AgCl-NPs

Polysaccharide-based bioflocculant reduction of Ag^+ ions, formation of AgCl-NPs/AgNPs in aqueous solution was monitored for 12 h, fivedays, 10 days and twomonths by measuring the ultraviolet-visible absorbance spectrum of the solution using a UV-visible spectrophotometer (TU-1901, Beijing Purkinje General Instrument Co., Ltd., Beijing, China) in the range of 200–700 nm [24]. Scanning electronmicroscope EDX analysis of the AgCl-NPs was performed using a Hitachi apparatus, Japan. The phase composition and crystal structure of the AgCl-NPs was determined using XRD (Bruker D8 advance, Karlsruhe, Germany). For this, the dried sample was prepared by placing on the microscopic glass slide and the diffractogram was recorded using Cu-K α radiation and a nickel monochromator filtering wave at a voltage and current of 40 kV and 30 mA, respectively. The FT-IR spectrum of the polysaccharide-based bioflocculant-stabilized AgCl-NPs were analyzed using FT-IR spectroscopy (JASCO FT-IR 460, Daejon, Korea) operated at resolution of 4 cm^{-1}. For the measurement of FT-IR spectrum, the dried sample was powdered by grinding with KBr pellets and pressed into a mold. The spectrum was recorded at a frequency range of 500–4000 cm^{-1} [24].

4.6. Antibacterial and Antifungal Activities of AgCl-NPs

4.6.1. Incubation of Pathogenic Strains

All bacterial strains were cultured following manufacturers' culturing guidelines. Typically, *E. coli* ATCC11229 and *S. aureus* ATCC6538 strains were cultured in Luria-Bertani (LB) culture medium (tryptone, 10 g/L; yeast extract, 5 g/L; NaCl, 10 g/L) at 37 °C. In all the experiments, the concentrations of bacteria were determined by optical density at 600 nm.

C. albicans ATCC10231 was cultured in culture medium comprising of: glucose, 40 g/L and peptone, 10 g/L [25,39].

4.6.2. The Hole Method

The antibacterial and antifungal activity of the polysaccharide-based bioflocculant stabilized AgCl-NPs was measured using the agar well diffusion method. Bacterial and fungal pathogens such as *E. coli*, *S. aureus* and *C. albicans* were used as indicator strains for this analysis. The bacterial and fungal strains were aseptically inoculated into respective broth, and then incubated at 37 °C. Samples from the culture liquids of respective pathogen plated on Petri plates and wells were made using an agar well borer. Different concentrations of AgCl-NPs were added to these wells, and the plates were incubated at 37 °C for 24 h. Zone of inhibitions were estimated by measuring the diameter of the bacterial growth inhibition zone. The values were averaged from the three independent experiments [25,39].

5. Conclusions

In conclusion, the polysaccharide-based bioflocculant of *A. chroococcum* XU1 can reduce Ag$^+$ from AgNO$_3$, and synthesize AgCl-NPs with absorption maximum at 250–260 nm. It was found that formation of AgCl-NPs and AgNPs was concentration dependent. At equal molar concentrations of Ag$^+$ and Cl$^-$ of the polysaccharide-based bioflocculant act as a template to assembly only AgCl-NPs with an average diameter 10–68 nm. XRD analysis of AgCl-NPs exhibited characteristic peaks at 27.64°, 32.24°, 46.2°, 54.78°, 57.44°, 67.42°, 74.4° and 76.6° can be attributed to the (111), (200), (220), (311), (222), (400), (331) and (420) of cubic phase of AgCl crystal. When Ag$^+$ concentration exceeded than that of Cl$^-$ formation of AgCl-NPs/AgNPs observed. It is supposed that in the presence of Cl$^-$ ions, synthesis of AgCl and its assembly over the polysaccharide-based bioflocculant occurs first. When the chloride ions depleted in the reaction mixture, excess amount of silver ions are reduced and stabilized in the polysaccharide-based bioflocculant, leading the formation of AgCl-NPs/AgNPs. As-synthesized AgCl-NPs exhibited strong antibacterial and antifungal activity against *E. coli*, *S. aureus* and *C. albicans*.

Acknowledgments: This work was supported by the Projects of International Cooperation and Exchanges of the National Natural Science Foundation of China (Grant No. 31110103908), the Projects of International Science & Technology Cooperation of the Xinjiang Uyghur Autonomous Region (Grant No. 20126023), and the Central Asian Drug Discovery and Development Center of Chinese Academy of Sciences.

Author Contributions: Experimental design was planned by Bakhtiyor A. Rasulov. Bakhtiyor A. Rasulov, Parhatjan Rozi, Mohichehra A. Pattaeva and Abulimiti Yili did the experimental work. Haji Akber Aisa supervised the project and revised the manuscript. Bakhtiyor A. Rasulov wrote the manuscript.

Conflicts of Interest: The authors declare no conflict of interest.

References

1. Manivasagan, P.; Kang, K.-H.; Kim, D.G.; Kim, S.-K. Production of polysaccharide-based bioflocculant for the synthesis of silver nanoparticles by *Streptomyces* sp. *Int. J. Biol. Macromol.* **2015**, *77*, 159–167. [CrossRef] [PubMed]
2. Lian, B.; Chen, Y.; Zhao, J.; Teng, H.H.; Zhu, L.; Yuan, S. Microbial flocculation by *Bacillus mucilaginosus*: Applications and mechanisms. *Bioresour. Technol.* **2008**, *99*, 4825–4831. [CrossRef] [PubMed]
3. Nwodo, U.U.; Green, E.; Mabinya, L.V.; Okaiyeto, K.; Rumbold, K.; Obi, L.C.; Okoh, A.I. Bioflocculant production by a consortium of *Streptomyces* and *Cellulomonas* species and media optimization via surface response model. *Colloids Surf. B* **2014**, *116*, 257–264. [CrossRef] [PubMed]
4. Surendhiran, D.; Vijay, M.J. Influence of bioflocculation parameters of harvesting *Chlorella salina* and its optimization using response surface methodology. *Environ. Chem. Eng.* **2013**, *1*, 1051–1056. [CrossRef]
5. Aljuboori, A.H.R.; Idris, A.; Abdullah, N.; Mohamad, R. Production and characterization of a bioflocculant produced by *Aspergillus flavus*. *Bioresour. Technol.* **2013**, *127*, 489–493. [CrossRef] [PubMed]
6. Nahvi, I.; Emtiazi, G.; Alkabi, L. Isolation of a flocculating *Saccharomyces cerevisiae* and investigation of its performance in the fermentation of beet molasses to ethanol. *Biomass Bioenergy* **2002**, *23*, 481–486. [CrossRef]
7. Shih, I.; Van, Y.; Yeh, L.; Lin, H.; Chang, Y. Production of a biopolymer flocculant from *Bacillus licheniformis* and its flocculation properties. *Bioresour. Technol.* **2001**, *78*, 267–272. [CrossRef]
8. Ye, S.; Ma, Z.; Liu, Z.; Liu, Y.; Zhang, M.; Wang, J. Effects of carbohydrate sources on biosorption properties of the novel exopolysaccharides produced by *Arthrobacter* ps-5. *Carbohydr. Polym.* **2014**, *112*, 615–621. [CrossRef] [PubMed]
9. Rahul, R.; Jha, U.; Sen, G.; Mishra, S. Carboxymethyl inulin: A novel flocculant for wastewater treatment. *Int. J. Biol. Macromol.* **2014**, *63*, 1–7. [CrossRef] [PubMed]
10. Kolya, H.; Tripathy, T. Preparation, investigation of metal ion removal and flocculation performances of grafted hydroxyethyl starch. *Int. J. Biol. Macromol.* **2013**, *62*, 557–564. [CrossRef] [PubMed]
11. Kanmani, P.; Satish, K.R.; Yuvaraj, N.; Paari, K.A.; Pattukumar, V.; Arul, V. Probiotics and its functionally valuable products. *Crit. Rev. Food Sci. Nutr.* **2012**, *53*, 641–658. [CrossRef] [PubMed]

12. Santander-Ortega, M.J.; Fuente, M.D.L.; Lozano, M.V.; Bekheet, M.E.; Progatzky, F.; Elouzi, A. Hydration forces as a tool for the optimization of core-shellnanoparticle vectors for cancer gene therapy. *Soft Matter* **2012**, *8*, 12080–12092. [CrossRef]

13. Abbasi, A.R.; Morsali, A. Synthesis and characterization of AgCl nanoparticles under various solvents by ultrasound method. *J. Inorg. Oraganomet. Polym. Mater.* **2013**, *23*, 286–292. [CrossRef]

14. Khan, Z.; Al-Thabaiti, S.A.; Obaid, A.Y.; Al-Youbi, A.O. Preparation and characterization of silver nanoparticles by chemical reduction method. *Colloids Surf. B Biointerfaces* **2011**, *82*, 513–517. [CrossRef] [PubMed]

15. Zhang, W.; Qiao, X.; Chen, J. Synthesis and characterization of silver nanoparticles in AOT microemulsion system. *Chem. Phys.* **2006**, *30*, 495–500. [CrossRef]

16. Rodriguez-Sánchez, L.; Blanco, M.C.; Lopez-Quintela, M.A. Electrochemical synthesis of silver nanoparticles. *J. Phys. Chem. B* **2000**, *104*, 9683–9686. [CrossRef]

17. Grzelczak, M.; Liz-Marzan, L.M. The Relevance of light in the formation of colloidal metal nanoparticles. *Chem. Soc. Rev.* **2014**, *43*, 2089–2097. [CrossRef] [PubMed]

18. Wang, P.; Huang, B.; Qin, X.; Zhang, X.; Dai, Y.; Wei, J.; Whangbo, M. Ag@AgCl: A highly efficient and stable photocatalyst active under visible light. *Angew. Chem. Int. Ed.* **2008**, *47*, 7931–7933. [CrossRef] [PubMed]

19. An, C.; Peng, S.; Sun, Y. Facile synthesis of sunlight-driven AgCl:Ag plasmonic nanophotocatalyst. *Adv. Mater.* **2010**, *22*, 2570–2574. [CrossRef] [PubMed]

20. Liu, C.; Yang, D.; Wang, Y.; Shi, J.; Jiang, Z. Fabrication of antimicrobial bacterial cellulose-Ag/AgCl nanocomposite using bacteria as versatile biofactory. *J. Nanopart. Res.* **2012**, *14*, 1084. [CrossRef]

21. Lee, K.Y.; Won, T.P. Green synthesis and antimicrobial activity of silver chloride nanoparticles stabilized with chitosan oligomer. *J. Mater. Sci. Mater. Med.* **2014**, *12*, 26–29.

22. Hu, W.; Chen, S.; Li, X.; Shi, S.; Shen, W.; Zhang, X.; Wang, H. In situ synthesis of silver chloride nanoparticles into bacterial cellulose membranes. *Mater. Sci. Eng. C* **2009**, *29*, 1216–1219. [CrossRef]

23. Dong, Y.-Y.; Deng, F.; Zhao, J.-J.; He, J.; Ma, M.-G.; Xu, F.; Sun, R.-C. Environmentally friendly ultrasound synthesis and antibacterial activity of cellulose/Ag/AgCl hybrids. *Carbohydr. Polym.* **2014**, *99*, 166–172. [CrossRef] [PubMed]

24. Rasulov, B.A.; Pattaeva, M.A.; Yili, A.; Aisa, H.A. Polysaccharide-based bioflocculant template of a diazotrophic *Bradyrhizobium japonicum* 36 for controlled assembly of AgCl nanoparticles. *Int. J. Biol. Macromol.* **2016**, *89*, 682–688. [CrossRef] [PubMed]

25. Rasulov, B.A. Obtaining and activity of silver nanoparticles based on the exopolysaccharide of diazotrophic strain *Bradyrhizobium japonicum* 36 and AgNO₃. *Acta Biotechnol.* **2014**, *7*, 57–62. [CrossRef]

26. Kanmani, P.; Lim, S.T. Synthesis and structural characterization of silver nanoparticles using bacterial exopolysaccharide and its antimicrobial activity against food and multidrug resistant pathogens. *Process Biochem.* **2013**, *48*, 1099–1106. [CrossRef]

27. Wang, P.; Huang, B.; Lou, Z.; Zhang, X.; Qin, X.; Dai, Y.; Zheng, Z.; Wang, X. Synthesis of highly efficient Ag@AgCl plasmonic photocatalysts with various structures. *Chem. Eur. J.* **2010**, *16*, 538–544. [CrossRef] [PubMed]

28. Bankura, K.P.; Maity, D.; Mollick, M.M.R.; Mondal, D.; Bhowmick, B.; Bain, M.K. Synthesis, characterization and antimicrobial activity of dextran stabilized silver nanoparticles in aqueous medium. *Carbohydr. Polym.* **2012**, *89*, 1159–1165. [CrossRef] [PubMed]

29. Pandey, S.; Goswami, G.K.; Nanda, K.K. Green synthesis of biopolymer–silver nanoparticle nanocomposite: An optical sensor for ammonia detection. *Int. J. Biol. Macromol.* **2012**, *51*, 583–589. [CrossRef] [PubMed]

30. Cui, L.; Jiao, T.; Zhang, Q.; Zhou, J.; Peng, Q. Facile preparation of silver halide nanoparticles as visible light photocatalysts. *Nanomater. Nanotechnol.* **2015**, *5*, 20. [CrossRef]

31. Sharma, R.K.; Lalita. Synthesis and characterization of graft copolymers of N-Vinyl-2-Pyrrolidone onto guar gum for sorption of Fe²⁺ and Cr⁶⁺ ions. *Carbohydr. Polym.* **2011**, *83*, 1929–1936. [CrossRef]

32. Hassabo, A.G.; Nada, A.A.; Ibrahim, H.M.; Abou-Zeid, N.Y. Impregnation of silver nanoparticles into polysaccharide substrates and their properties. *Carbohydr. Polym.* **2014**, *120*, 343–350.

33. Shankar, S.; Rhim, J.W. Amino acid mediated synthesis of silver nanoparticles and preparation of antimicrobial agar/silver nanoparticles composite films. *Carbohydr. Polym.* **2015**, *130*, 353–363. [CrossRef] [PubMed]

34. Kora, A.J.; Arunachalam, J. Assessment of antibacterial activity of silver nanoparticles on *Pseudomonas aeruginosa* and its mechanism of action. *World J. Microbiol. Biotechnol.* **2011**, *27*, 1209–1216. [CrossRef]

35. Wei, X.; Luo, M.; Li, W.; Yang, L.; Liang, X.; Xu, L. Synthesis of silver nanoparticles by solar irradiation of cell-free *Bacillus amyloliquefaciens* extracts and AgNO₃. *Bioresour. Technol.* **2012**, *103*, 273–278. [CrossRef] [PubMed]

36. Panacek, A.; Kolar, M.; Vecerova, R.; Prucek, R.; Soukupova, J.; Krystof, V.L. Antifungal activity of silver nanoparticles against *Candida* spp. *Biomaterials* **2009**, *30*, 6333–6340. [CrossRef] [PubMed]

37. Durán, N.; Cuevas, R.; Cordi, L.; Rubilar, O.; Diez, M.C. Biogenic silver nanoparticles associated with silver chloride nanoparticles (Ag@AgCl) produced by laccase from *Trametes versicolor*. *SpringerPlus* **2014**, *3*, 645. [CrossRef] [PubMed]

38. Awwad, A.M.; Salem, N.M.; Ibrahim, Q.M.; Abdeen, A.O. Phytochemical fabrication and characterization of silver/silver chloride nanoparticles using *Albizia julibrissin* flowers extract. *Adv. Mater. Lett.* **2015**, *6*, 726–730.

39. Rasulov, B.A.; Rustamova, N.; Qing, Z.H.; Yili, A.; Aisa, H.A. Synthesis of silver nanoparticles on the basis of low- and high-molar-mass exopolysaccharides of *Bradyrhizobium japonicum* 36 and its antimicrobial activity against some pathogens. *Folia Microbiol.* **2016**, *61*, 283–293. [CrossRef] [PubMed]

![materials logo] *materials*

MDPI

Article

Synthesis and Characterization of New Chlorhexidine-Containing Nanoparticles for Root Canal Disinfection

Ridwan Haseeb [1], Michael Lau [1], Max Sheah [1], Francisco Montagner [2,*], Gina Quiram [1], Kelli Palmer [3], Mihaela C. Stefan [1,4] and Danieli C. Rodrigues [1,*]

[1] Department of Bioengineering, University of Texas at Dallas, 800 W Campbell, Richardson, TX 75080, USA; rbh110020@utdallas.edu (R.H.); mxl110330@utdallas.edu (M.L.); mms097020@utdallas.edu (M.S.); gjq140030@utdallas.edu (G.Q.); mihaela@utdallas.edu (M.C.S.)
[2] Department of Conservative Dentistry, Dental School, Federal University of Rio Grande do Sul, Rua Ramiro Barcelos 2492, Porto Alegre–RS 90460-001, Brazil
[3] Department of Molecular and Cell Biology, University of Texas at Dallas, 800 W Campbell, Richardson, TX 75080, USA; Kelli.Palmer@utdallas.edu
[4] Department of Chemistry, University of Texas at Dallas, 800 W Campbell, Richardson, TX 75080, USA
* Correspondence: francisco.montagner@ufrgs.br (F.M.); dxb127430@utdallas.edu (D.C.R.); Tel.: +1-972-883-4703 (D.C.R.)

Academic Editor: Mauro Pollini
Received: 20 April 2016; Accepted: 30 May 2016; Published: 7 June 2016

Abstract: Root canal system disinfection is limited due to anatomical complexities. Better delivery systems of antimicrobial agents are needed to ensure efficient bacteria eradication. The purpose of this study was to design chlorhexidine-containing nanoparticles that could steadily release the drug. The drug chlorhexidine was encapsulated in poly(ethylene glycol)–*block*–poly(L-lactide) (PEG–*b*–PLA) to synthesize bilayer nanoparticles. The encapsulation efficiency was determined through thermogravimetric analysis (TGA), and particle characterization was performed through microscopy studies of particle morphology and size. Their antimicrobial effect was assessed over the endodontic pathogen *Enterococcus faecalis*. The nanoparticles ranged in size from 300–500 nm, which is considered small enough for penetration inside small dentin tubules. The nanoparticles were dispersed in a hydrogel matrix carrier system composed of 1% hydroxyethyl cellulose, and this hydrogel system was observed to have enhanced bacterial inhibition over longer periods of time. Chlorhexidine-containing nanoparticles demonstrate potential as a drug carrier for root canal procedures. Their size and rate of release may allow for sustained inhibition of bacteria in the root canal system.

Keywords: dentin tubules; dentin permeability; chlorhexidine; nanoparticles; encapsulation

1. Introduction

The treatment of an infected root canal has been based on nonspecific elimination of intraradicular microorganisms through the application of broad-spectrum antimicrobial approaches [1]. Nevertheless, it has been shown that it is nearly impossible to obtain complete elimination of microorganisms in the root canal system [2,3]. Therefore, the continued development of treatments that can effectively penetrate dentin to eliminate root canal infection is a priority in clinical endodontic research.

Currently, the most frequent intracanal medicaments employed to treat infected root canals include calcium hydroxide ($Ca(OH)_2$), potassium iodine (KI), and chlorhexidine (CHX). The efficacy of root canal disinfectants can be influenced by several factors such as pH, serum proteins, collagen, and dentin among others [4–6]. However, *in vitro* studies have demonstrated that CHX is more effective in eliminating bacteria from internal dentinal tubules in comparison to the other disinfectants when dispersed in liquid or hydrogel systems [7,8]. Besides its proven antimicrobial activity and non-toxicity at low dosages, CHX provides substantivity to dentin tissues, which may offer protection against microbial colonization for extended periods of time after treatment [7,9]. These attributes make CHX a potent disinfectant in root canal treatment. Dentin permeability and the complex anatomy of the root canal, however, impose challenges to the penetration and subsequent action of these disinfectants. Gomes *et al.* [10] demonstrated *in vitro* that medicaments containing 2% CHX were able to diffuse into the dentin, reaching the external root surface. The information on the length of activity time of the various agents in the root canal is limited.

Nanoparticles in recent years have been employed in several clinical applications [11]. In endodontics, nanoparticles have been suggested to act as irrigants [12], incorporated into intracanal medicaments [13] or root canal sealers [14,15]. Nanoparticle technology for drug delivery includes nanoencapsulation, which is the coating of a substance within another material, typically a polymer based system. It aims to maximize the therapeutic efficacy while minimizing undesirable side effects due to the control of the drug bioavailability and release [16,17]. There is scarce data in the current literature on the synthesis, characterization, and application of nanoencapsulated medicaments that are typically employed in root canal treatment. Shrestha and Kishen [18] evaluated the effect of rose bengal-functionalized chitosan nanoparticles associated with photodynamic therapy over monospecies bacteria/biofilms and assessed their antibiofilm efficacy on a multispecies biofilm grown on dentin. Shrestha *et al.* [19] examined the ability of the temporally controlled release of bovine serum albumin from chitosan nanoparticles to regulate the alkaline phosphatase activity in stem cells from apical papilla.

Nanoparticles produced by drug nanoencapsulation have specific characteristics such as size, release pattern, and activity, which are factors determined by the synthetic method employed, polymer system of choice, and polymer molecular weight. Therefore, protocols should be conducted to achieve proper nanoencapsulation of medicaments, taking into account the stability of the system or release profile to achieve the desirable antiseptic activity inside a specific target or tissue. Provided the anatomical complexity of the root canal, permeability of dentin, and limited penetration of medicaments in the dentin, the goal of this study is to develop and characterize a novel CHX-encapsulated system for root canal applications. In this work, the biodegradable and biocompatible block copolymer of choice was poly(ethylene glycol)–*block*–poly(L-lactide) (PEG–*b*–PLA) to create CHX loaded nanoparticles. PEG–*b*–PLA bilayer nanoparticles have advantages for drug delivery, such as small size and hydrophobic and hydrophilic functionalities in the polymer backbone that improves *in vivo* half-life. These polymeric nanoparticles were characterized for size, morphology, and drug loading proficiency. The nanoparticles were found to be small enough to penetrate dentin tubules, dispersed well in a hydrogel matrix used as a carrier system, and enhanced bacterial inhibition over longer periods of time.

2. Results

2.1. Particle Synthesis

The obtained PEG–*b*–PLA block copolymer was characterized by ^{1}H NMR. Figure 1 shows the ^{1}H NMR spectra of the PEG–*b*–PLA block copolymer and indicates that the block copolymer was synthesized successfully. The number-average molecular weight of the PEG–*b*–PLA block copolymer was 5756 daltons with a ratio of 2 PLA units to 1 PEG unit. The peaks at 3.64 ppm and 3.38 ppm corresponded to methylene units and CH_3O- in the mPEG block, respectively. Signals at 1.47 ppm and 5.16 ppm could be attributed to the hydrogen atoms of CH_3- and $CH-$ groups for PLA segments, respectively.

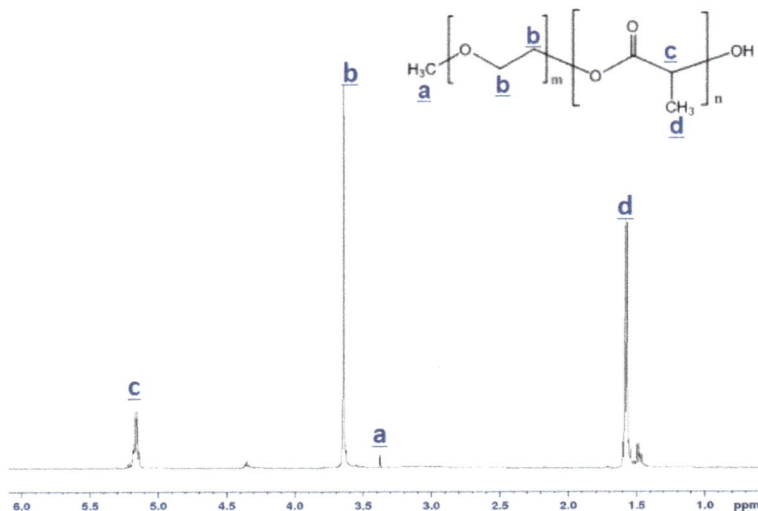

Figure 1. ^{1}H NMR spectra of the synthesized poly(ethylene glycol)–*block*–poly(L-lactide) (PEG–*b*–PLA) block copolymer.

2.2. Encapsulation Efficiency

The synthesis employed proved to be a simple and reproducible method to encapsulate CHX for controlled release of CHX. Investigation of the thermal behavior of synthesized products using thermogravimetric analysis (TGA) illustrated that the encapsulation of CHX was achieved. The measured mass drop of the materials as they decomposed using TGA (Figure 2) illustrates that the encapsulation process resulted in structural changes with the incorporation of CHX in PEG–*b*–PLA, as there is a slight shift in the decomposition point of the materials. TGA also provided information on the encapsulation efficiency of CHX in PEG–*b*–PLA.

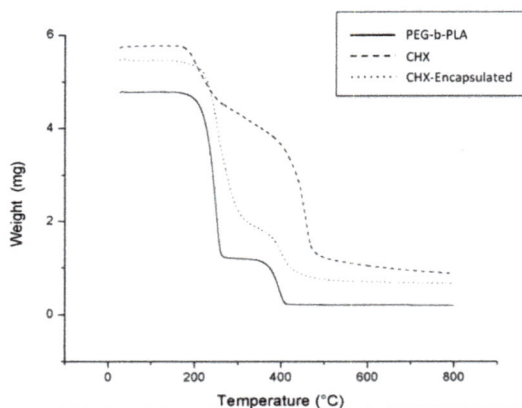

Figure 2. Measurements of the amount of weight change of PEG–*b*–PLA, chlorhexidine (CHX), and CHX-encapsulated. Thermogravimetric analysis (TGA) curves showing mass loss as a function of temperature. Note that thermal transitions for the CHX-encapsulated product most resemble the behavior and transitions measured for the PEG–*b*–PLA polymer due to the proportion of PEG–*b*–PLA being much higher than CHX.

The synthetic method employed allowed for an average encapsulation efficiency of 70%. Energy dispersive X-ray spectroscopy (EDS) analysis confirmed the presence of CHX in the encapsulated nanoparticles by signal emission of chlorine (Cl) at 2.7 keV (Figure 3).

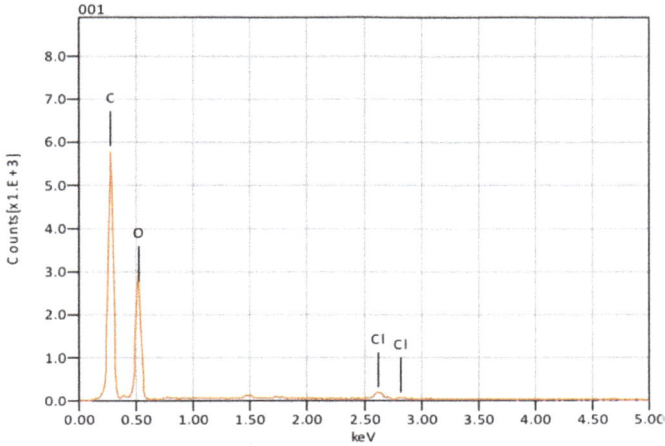

Figure 3. Nanoparticle composition obtained with energy dispersive X-ray spectroscopy (EDS).

2.3. Particle Morphology and Composition

Scanning electron microscopy (SEM) at 1800× magnification (Figure 4a) was used to verify the physical characteristics of the nanoparticles and presence of clumps. The analysis showed that CHX-encapsulated formed large clumps in the dry state, which prevented accurate analysis using this technique. Atomic force microscopy (AFM) was therefore used to determine size AFM revealed that nanoparticles individually ranged in size from 300–500 nm in diameter (Figure 4b,c). The size was confirmed by probing numerous areas on the glass slides containing the nanoparticles using topographical and 3D imaging. From dynamic light scattering (DLS) analysis, the nanoparticle's average size was found to be 342 nm.

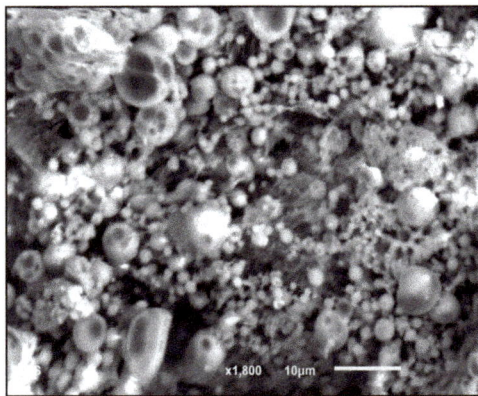

(a)

Figure 4. *Cont.*

(b)

(c)

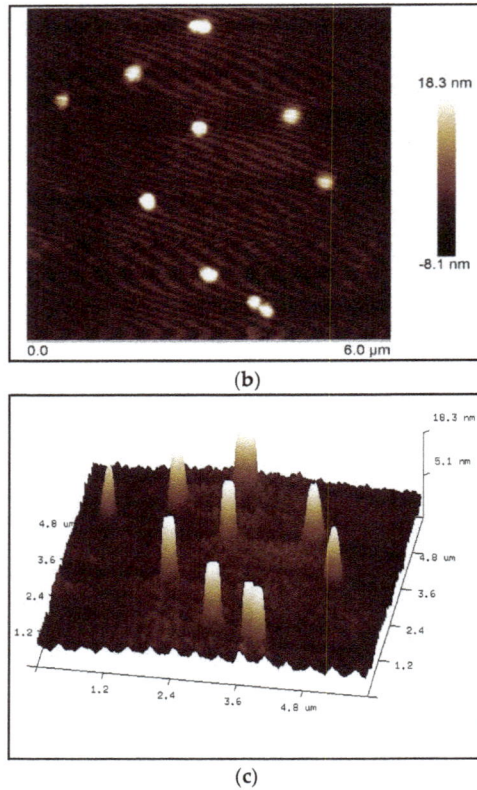

Figure 4. (a) Morphological analysis obtained with scanning electron microscopy (SEM) images of CHX-encapsulated nanoparticles. Note that clumping prevented measurement of individual nanoparticle size. Atomic force microscopy (AFM) (b) 2D surface topography of nanoparticles and (c) 3D image.

2.4. Antimicrobial Effectiveness

The ability of the CHX-encapsulated nanoparticles to retain and release CHX was investigated with zone of inhibition (ZOI) measurements. The ZOI generally became smaller as the filtered nanoparticles that spent greater time immersed in phosphate buffer saline (PBS) were placed on bacterial lawns. Control nanoparticles (synthesized without the addition of CHX) showed no ZOI, indicating that the polymer itself did not have antimicrobial activity. Table 1 shows the proportion of nanoparticle mass diameter compared to the ZOI. The table shows that nanoparticles immersed for 14 days and those immersed for 21 days displayed similar ZOIs.

Table 1. Zone of inhibitions measured for CHX-encapsulated nanoparticles filtered from phosphate buffer saline (PBS). The values indicate the results obtained in three trials averaged together.

Days Particles Were Immersed in PBS	Ratio of Assembled Nanoparticles Diameter *vs.* Zone Diameter
1 h	0.436 ± 0.042
7	0.584 ± 0.019
14	0.659 ± 0.042
21	0.675 ± 0.043

The optical density (OD) data from filtered bacterial broth that had contained the nanoparticles for 7, 14 and 21 days obtained through the plate reader showed the antimicrobial effect displayed by the CHX-encapsulated nanoparticles. The onset of *E. faecalis* exponential growth was delayed by approximately three hours for nanoparticle immersion solutions as compared to the control broth, and the final OD was slightly lower in the broths that had contained the CHX nanoparticles *vs.* the control. The OD data at the intermediate time period when the control was entering the growth phase was also analyzed using a one-way ANOVA. When the growth curves of the broth that had contained the nanoparticles were compared to the control, a significant difference was observed ($P \leqslant 0.05$). The results also indicated that the nanoparticle mass remained effective for the period investigated (up to 21 days). However, the final ODs as well as the lag for the growth phases remained nearly the same (Figure 5).

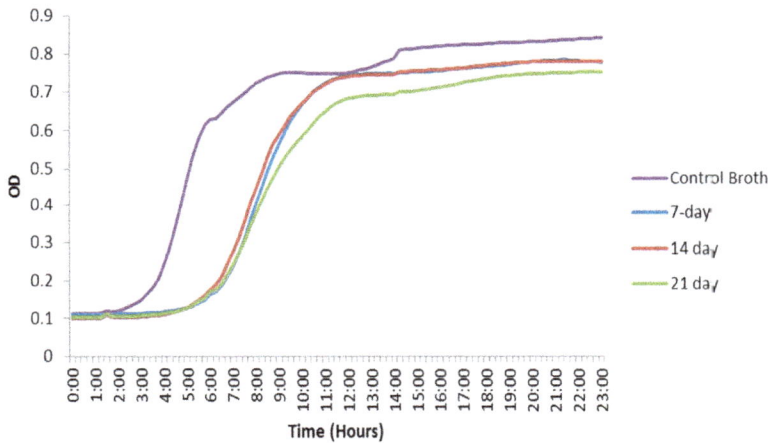

Figure 5. Optical density curves of *E. faecalis*. Broth that had contained CHX-encapsulated nanoparticles show a growth delay compared to the positive control broth that had not contained any nanoparticles.

3. Discussion

The present study discussed the synthesis, characterization, and antimicrobial effectiveness of a new drug delivery system designed to allow for extended release of CHX inside the root canal system. It was hypothesized that nanoparticles prepared with an appropriate size and a controlled CHX-release profile could carry the medicament deep into dentin tubules, allowing for sustained inhibition of bacteria in the root canal system.

Chlorhexidine has been used previously in hydrogel form and in liquid formulations in root canal treatment. CHX is well-known to be rapidly released from poly(lactic acid) microparticles [20]. Therefore, in order to enhance bacterial elimination from infected root canal systems and to create an environment that is ideal for periapical healing, it is highly desirable to extend the release period of CHX *in situ* beyond the delivery systems currently in use. Nanoencapsulation can provide better control while further extending the release period of the medicament in dentin tissues. The antibacterial activity of CHX in different concentrations and preparation forms has been extensively tested. In endodontics, CHX has been employed as an intracanal medicament in a 2% concentration, alone or associated with calcium hydroxide [21]. It should also be emphasized that, the higher the drug concentration, the higher its side effects.

The bacterial activity of CHX loaded poly(ε-caprolactone) nanocapsules and CHX digluconate against *S. epidermis* was studied in [22]. The CHX carrier system was observed to improve drug targeting of bacteria, further reducing bacterial growth onto skin in relation to CHX digluconate. In the present study, CHX was encapsulated in a poly(ethylene glycol)–*block*–poly(L-lactide) ("PEG–*b*–PLA"). The choice of PEG–*b*–PLA is optimal as poly(L-lactic acid) has long been known for its biocompatibility and biodegradation properties [23]. It is a common choice to safely encapsulate bioactive drugs and control their release [24]. PEG has been copolymerized with PLA to create a block polymer and facilitate drug release due to its dual nature, hydrophilic-hydrophobic, which can increase pore formation in the nanoparticles along with a rise in the rate of polymer degradation. PLA microspheres degrade through a hydrolytic chain cleavage reaction affecting both the surface and the bulk properties of spheres while exhibiting no serious health risks [24]. All these properties make the copolymer a viable candidate to both encapsulate and release CHX. The PEG–*b*–PLA nanoparticles were prepared through the oil-in-water emulsion technique. The goal of this methodology was to synthesize nanoparticles smaller than 1 μm in order to facilitate penetration of the nanoparticles deep into the tubules. The encapsulation process resulted in nanoparticles with a size ranging much smaller than the diameter of dentin tubules [25]. Atomic force microscopy revealed that the size of the nanoparticles ranged from 300 to 500 nm (Figure 4), which was verified by dynamic light scattering (DLS).

The PEG–*b*–PLA nanoparticles prepared through the oil-in-water emulsion technique resulted in high material yields and drug encapsulation efficiencies as high as 80%. TGA revealed that CHX was incorporated into the polymer as demonstrated by weight changes of decomposition reactions of the CHX-containing nanoparticles in comparison to starting materials—CHX and the PEG–*b*–PLA block copolymer (Figure 2).

The presence of chlorine peaks (Figure 3) in the EDS spectra are evidence of CHX-encapsulated nanoparticles. As this element is characteristic of CHX and not the polymer, there is evidence to suggest that the CHX is being absorbed on the surface of the nanoparticles and encapsulated. Although SEM showed that the nanoparticles can form large clumps during the lyophilization process, these were significantly reduced when the nanoparticles were dispersed in a hydrogel matrix made out of 1% Natrosol™ (hydroxyethyl cellulose) in water.

Nanoparticle disperssion within the hydrogel matrix consists of polymer chains from the hydrogel forming weak bonds with many nanoparticles. This in turn produces a loosely interlinked network of polymers and nanoparticles [26]. Since each connection point is rather weak, the bonds breakdown under mechanical stress, for instance, when injected through syringes [27]. When the shear forces subside, the polymers and nanoparticles form new connections with the walls of dentinal tubules and within the hydrogel itself. Now, the high water content and large pore sizes of this hydrogel will be sufficient to trigger hydrolysis of the backbone ester groups in PLA and the oxidation of the ether backbone in PEG, thereupon diffusing the chlorhexidine.

Microorganisms may penetrate inside dentin to different extents and may survive inside the tubules, even after the use of the currently employed disinfection protocols. In a clinical study, Siqueira *et al.* [28] demonstrated that chemomechanical preparations with 2.5% NaOCl as irrigant significantly reduced the number of bacteria in the canal, but allowed for microbial recovery through cultivation in more than one-half of all cases. Authors also reported that a seven-day intracanal dressing with $Ca(OH)_2$/camphorated paramonochlorophenol (CPMC) paste significantly increased the number of culture-negative cases. However, positive cultures were still obtained. Furthermore, *Enterococcus faecalis*, a facultative anaerobe that can be isolated from persistent/secondary root canal infections, has demonstrated to be resistant to calcium hydroxide, especially when in biofilms. They can adapt to harsh environmental changes and can colonize dentinal tubules where they remain protected from medicaments. Upadya *et al.* [29] showed that *E. faecalis* biofilms were considerably more resistant to $Ca(OH)_2$ solutions than free-floating cells. A fraction from the biofilm cells persisted viable even after 24-h exposure to a saturated $Ca(OH)_2$ solution. It was pointed out that the increased resistance might be attributed to the biofilm structure or extrapolimeric substance. Therefore, other antimicrobial agents, with different mechanisms of action should be employed to enhance the root canal system disinfection, especially in areas that are difficult to reach with conventional instrumentation and currently employed intracanal medicament. Chlorhexidine digluconate is water-soluble and readily dissocates at physiologic pH releasing the positively charged chlorhexidine component. The bactericidal effect of the drug is due to the cationic molecule binding to extra microbial complexes and negatively charged microbial cell walls, thereby altering the cells' osmotic equilibrium [30,31]. Chlorhexidine with a concentration of 2%, in both liquid or gel presentations, has been employed as an auxiliary chemical substance during root canal preparation or as a final irrigant [32–34]. Several *in vitro* studies assessed the properties of CHX as an intracanal medicament [10,35,36]. However, there are few clinical studies that employed CHX hydrogel as an intracanal medicament. Gama *et al.* [37] evaluated the incidence of postoperative pain after intracanal dressings with either 0.12% chlorhexidine digluconate gel or a calcium hydroxide/camphorated paramonochlorophenol/glycerin paste. Therefore, this study intended to develop a nanoparticle–hydrogel matrix system to be employed as an intracanal medicament that would carry CHX, allowing for its continuous release inside the root canal system.

Few studies have assessed the antimicrobial effect of antimicrobial-containing nanoparticles over endodontic pathogens, especially *Enterococcus faecalis*. The effect of rose bengal-functionalized chitosan nanoparticles over *E. faecalis* cells and multispecies biofilm structures was assessed by Shrestha and Kishen [18]. Despite the quantification of the CHX release from nanoparticles assessed through analytical methods [38–40], no study has evaluated the residual antimicrobial effect produced by suspended nanoparticles over time, especially against *E. faecalis* cells. In the present study, *in vitro* studies in *E. faecalis* were performed to investigate bacterial growth inhibition in the presence of CHX-encapsulated nanoparticles. The experiment was performed at different time points to investigate the length of antibacterial activity and effectiveness of the synthesized delivery system. The initial goal was to maintain CHX release for a minimum period of 7 to 14 days, which corresponds to the period that an intracanal medicament (such as calcium hydroxide paste) is often kept between appointments. The synthesized nanoparticles were immersed in both PBS and BHI (brain–heart infusion) broth. Nanoparticles removed from the PBS and placed on bacterial lawns showed a zone of inhibition after being immersed for as long as 21 days. These zones of inhibition demonstrate that CHX both elutes into the solution and remains in the particles after eluting for up to three weeks. They also provide evidence of the area where bacteria will not come into contact with the material even after eluting the drug into a solution. The broth containing the nanoparticles demonstrated a lag in the growth phase of *E. faecalis* bacteria as well as a decrease in total cell density during this lag phase, indicating inhibition and demonstrating that CHX was indeed diffusing from the nanoparticles. The diffusion of CHX is directed by chemical potential gradients arising by osmotic pressure. In addition to diffusion, CHX could be released by erosion of the polymer matrix, which leads to pore formation [41].

The initial burst release behavior of the nanoparticles could be attributed to the CHX absorbed on the surface of the nanoparticles, with subsequent releases corresponding to the encapsulated portion. Both results (ZOI and OD) suggest that the nanoparticles demonstrate a bacteriostatic effect for at least three weeks. The zone of inhibition tests showed that CHX was eluting from the hydrophobic core of the PEG–*b*–PLA nanoparticles into the PBS solution, while some CHX still remained in the nanoparticles and were still capable of antimicrobial action. Furthermore, the bacterial growth curves using the OD data demonstrated that eluting CHX into bacterial broth did have an effect on the *E. faecalis* bacteria's growth curve. However, the size of the zones of inhibition decreased over time. Eventually, the release of CHX reached equilibrium, and the nanoparticles that had been immersed for 14 days in PBS showed similar-sized zones as those immersed for 21 days. This suggests that hydrolysis or enzymatic cleavage of the PEG–*b*–PLA backbone reached the hydrophobic core of the nanoparticle, which causes bulk erosion, hence releasing the remaining CHX during the time period between 14 and 21 days. Additionally, the bacterial growth curves in the broth eventually reached their growth phase after the initial lag and had a final OD similar to the control, albeit slightly lower.

The intent of this nanoparticle-hydrogel matrix system was to develop an intracanal medicament that allows for the distribution of the nanoparticles through the lateral tubules and accessory canals that are exposed during the root canal procedure. The nanoparticle–hydrogel matrix system should not remain in the root canal itself after the endodontist does adequate irrigations prior to canal filling. There have been studies on hydrogel formulations and their ability to insert into narrow and convoluted locations, such as in the inter-tubular dentin matrix [13,42,43]. Thus, we are using this adhesive property of hydrogels in order to deposit and fasten the nanoparticles to the walls of dentinal tubules and accessory canals.

Control PEG–*b*–PLA nanoparticles (synthesized without the addition of CHX) showed no ZOI, indicating that no reactive oxygen species (ROS) were introduced during the synthesis of the nanoparticles. ROS are partially reduced metabolites of oxygen, including hydrogen peroxide and hydroxyl radical, which can result from polymer degradation [44]; this in turn causes additional oxidative stress in many pathological pathways making them toxic to cells [45]. Free radical generation in the cell culture containing the nanoparticles can also arise from the cellular uptake of low molecular weight polymer chains that result from their degradation [46]. This likewise leads to cytotoxicity due to the stimulation of ROS and/or the accumulation of polymer degradation products inside the cell. During the 21-day antimicrobial effectiveness study of PEG–*b*–PLA nanoparticles synthesized without the addition of CHX, no inhibitory diameter was detected, indicating that, even if the polymer is degrading, cells are not being affected by its degradation products or triggering ROS production inside cells. The results from the control in this study are of significant importance in the design of nano-carrier systems for endodontic drug delivery, thus validating the well-known biodegradability and biocompatibility of nanoparticles made out of PEG–*b*–PLA and making this a safe and suitable system of CHX delivery inside dentinal tubules.

The physiochemical properties of the PEG–*b*–PLA nanoparticles will need to be further tuned in order to enhance bacterial inhibition over longer periods of time. This will allow for prolonged release in the dentin tubules to better ensure the success of a root canal procedure. Future studies will include increasing the concentration of nanoparticles in medium and assessing the nanoparticle hydrogel matrix delivery network inside the root canal system.

4. Materials and Methods

4.1. Materials

Poly(ethylene glycol) methyl ether [average molecular weight (Mn) = 2000 g/mol], L-lactide, toluene, tin (II) 2-ethylhexanoate, chlorhexidine, poly(vinyl alcohol) (PVA) [87%–90% hydrolyzed, average Mw = 30,000–70,000 g/mol], dichloromethane (DCM), 1x phosphate buffer saline (PBS), and pentane were used as received without further purification. All chemicals were purchased from Sigma-Aldrich (St. Louis, MO, USA).

4.2. Nanoparticles Synthesis

The process of preparing bilayer nanoparticles comprised two steps: (1) polymer synthesis followed by (2) encapsulation of the drug.

4.2.1. Polymer Synthesis

Poly(ethylene glycol)–*block*–poly(L-lactide) (PEG–*b*–PLA) was synthesized in house by introducing poly(ethylene glycol) methyl ether (Mn = 2000) and L-lactide (Lactide-(3*S*)-*cis*-3,6-Dimethyl-1,4-dioxane-2,5-diene) into toluene distilled over sodium benzyl phenol at 25 °C. The mixture was subsequently placed under vacuum to remove moisture. Tin (II) 2-ethylhexanoate was added as a catalyst during continuous stirring at 100 °C for 5 h. The resulting polymer was precipitated in pentane and air-dried at room temperature overnight. The chemicals used were purchased from Sigma-Aldrich (St. Louis, MO, USA). The number-average molecular weight of the block copolymer was determined by ^{1}H NMR (ADVANCE III 500 MHz, Bruker, Santa Barbara, CA, USA). Samples temperature was regulated for all measurements and was set at 25 °C.

4.2.2. Encapsulation Process

An oil-water-emulsion-evaporation method was carried out for encapsulation of chlorhexidine (CHX). The oil/organic phase consisted of PEG–*b*–PLA and CHX in a 5:1 ratio, respectively, dissolved in dichloromethane (DCM). The water phase was a 1% *w/v* solution of poly(vinyl alcohol) and deionized water. The organic and water phases were combined and emulsified using ultra-sonication for 1 min (Branson Ultrasonics Corporation, Bransonic CPX3800H, Danbury, CT, USA). The resulting emulsion was stirred for 2 h at 25 °C at atmospheric pressure to allow the organic solvent to evaporate. Finally, the particles were centrifuged, washed, and freeze-dried.

4.3. Characterization

The starting and final products were characterized using multiple techniques: (1) encapsulation efficiency and (2) microscopy studies of nanoparticle morphology and size.

4.4. Encapsulation Efficiency

Encapsulation efficiency was found using quantitative measurement of mass change from dehydration, decomposition, and oxidation with time and temperature of the initial polymer (PEG–*b*–PLA), CHX, and CHX-encapsulated nanoparticles. Mass changes, from their physiochemical reactions, were measured using thermogravimetric analysis (TGA, Metler Toledo TGA1, Greifensee, Switzerland). This instrument detects changes in weight that occur in a material with increasing temperature (25 °C to 800 °C). A heating rate of 5 °C per minute was applied with about 5 mg of sample used for each run. The temperatures at which the polymers and CHX characteristic mass dropped were recorded and compared. For the CHX curve, the temperature at which a significant mass loss percentage occurred was recorded (CHX_{Temp}) and used as a baseline temperature. As the percentage loss of CHX at this temperature was determined earlier, the percentage of CHX lost in the nanoparticles could be found as well. This percentage was used to find the amount of CHX in the encapsulated nanoparticles and subsequently used to find encapsulated efficiency of the polymer (Equation (1)). The polymer curve was used to ensure that the polymer mass loss temperature did not overlap with CHX_{Temp}. CHX was detected when characteristic changes in the thermal decomposition temperatures were observed for the CHX-encapsulated polymer in comparison to the starting materials (pure PEG–*b*–PLA and pure CHX).

$$Encapsulation\ Efficiency = \frac{Mass\ of\ CHX\ in\ Particles}{Mass\ of\ CHX\ used\ in\ Synthesis}. \tag{1}$$

4.5. Particle Morphology and Composition

The morphology and size of polymer and encapsulated nanoparticles were observed with scanning electron microscopy (SEM, JEOL JSM-6010LA, Peabody, MA, USA) and atomic force microscopy (AFM, Bruker, Bioscope Catalyst, Santa Barbara, CA, USA). For the SEM analysis, a thin layer of nanoparticles was deposited on a metallic stub. The composition of the materials was measured with energy dispersive X-ray spectroscopy (EDS, JEOL JSM-6010LA, Peabody, MA, USA), which enabled detection of the individual elements of the nanoparticles. For AFM analysis, CHX-encapsulated nanoparticles were dispersed in a hydrogel solution (1% Natrosol™ hydroxyethyl cellulose in water) and sonicated for about 30 h (Branson Ultrasonics Corporation, Bransonic CPX3800H, Danbury, CT, USA) to disrupt clumps. A drop of solution containing nanoparticles was deposited on a glass slide and was allowed to dry forming a thin film. The AFM analysis was performed by using the quantitative nanomechanics method (QNM) for compositional mapping, which, besides determining morphological features, enables quantitative measurement of nanoscale material properties. The nanoparticle's size and size distribution were also investigated by dynamic light scattering (DLS) technique, using a non-invasive backscatter optics (NIBS) (Zetasizer Nano ZS, Malvern, Worcestershire, UK), after suspending 2 mg of the nanoparticles in 10 mL of deionized water.

4.6. Antimicrobial Effectiveness

The antimicrobial effect of the encapsulated nanoparticles was tested against *Enterococcus faecalis* OG1RF. CHX-encapsulated nanoparticles were immersed in phosphate buffer saline (PBS) and agitated for set time periods (1 h, 7 days, 14 days and 21 days). Then, the nanoparticles were filtered from the solution using vacuum filtration and air-dried for approximately 24 h. *E. faecalis* overnight broth culture was spread onto BHI (brain-heart infusion) agar plates and ~1 mg of dried CHX-encapsulated nanoparticles filtered previously from the PBS were placed onto the bacterial lawns in a roughly circular formation to test for zones of growth inhibition (ZOI) around the nanoparticles. The bacterial lawn plates were incubated for 24 h at 37 °C, and the ZOIs were observed. These experiments were performed with nanoparticles that were immersed in a PBS solution for 1 h, 7 days, 14 days and 21 days to investigate the potency of bacterial inhibition and whether any drug burst release was present. Digital calipers, which are measurement tools, were used to determine the diameter of the ZOI and the diameter of a batch of nanoparticles placed on the plates. As the mass, shape, and area of nanoparticles placed on the plate were variable (nanoparticles charge made placement difficult on the plate surface), the relative diameter of an assembled group of nanoparticles was measured and compared to the diameter of the ZOI. Therefore, if the ratio increased, that meant the ZOI decreased, as this would signify that the zone was closer to the nanoparticle mass. Three trials of this test were performed.

CHX-encapsulated nanoparticles were also immersed in BHI broth for varied periods (7 days, 14 days and 21 days). These time periods represent how long the nanoparticles would be active inside dentinal tubules. The nanoparticles were filtered out, and the remaining broth was kept. The broth was then inoculated with *E. faecalis* at an initial optical density at 600 nm (OD_{600}) of ~0.001. Afterwards, 200 µL of the inoculated broth was aliquoted in triplicate into a 96-well plate and incubated at 37 °C for 24 h. The OD_{600} was monitored with a Monochromater-based Multi-Mode Microplate Reader (Synergy Mx, Winooksi, VT, USA) every 15 min during the 24-h incubation period. The OD data was used to generate bacterial growth curves made by averaging triplicates from three trials together, not including clear outliers obtained during the experiments. The mean values of intermediate OD readings when the control broth was entering the growth phase were analyzed statistically with one-way ANOVA (analysis of variance) method at a 5% significance level. The OD readings of the broth that had contained the CHX nanoparticles were compared to the control to verify that the broth with CHX nanoparticles was indeed causing a delay in the growth phase.

5. Conclusions

In this study, PEG–*b*–PLA bilayer nanoparticles for CHX delivery were successfully assembled to improve drug bioavailability and target drug delivery to dentin tubules. This synthesis allowed for the sustained inhibition of bacteria that could potentially be used in root canal systems. The bilayer polymeric nanoparticles employed featured a hydrophobic interior space that easily encapsulated CHX, a hydrophobic drug. CHX release was effective for up to 21 days with an initial burst, which may be attributed to the CHX absorbed on the surface of the nanoparticles and subsequently predominantly controlled by diffusion and degradation mechanisms. These results have potential implications for the design of CHX polymeric nanoparticles for the *in situ* treatments of the root canal.

Acknowledgments: The authors would like to thank the University of Texas at Dallas for providing resources for this research through startup funds (Danieli C. Rodrigues). NIH (1R21EB019175-01A1), Welch Foundation (AT-1740) (AT-1740), and NSF (DMR-15059550) is graetfully acknowledge (Mihaela C. Stefan). We thank the contributions of Izabelle M. Gindri, Lucas C. Rodriguez, and Shant Aghyarian in this study. We also thank the Department of Chemistry and Biochemistry at the University of Texas at Dallas for providing access to the NMR facilities (instrument grant NSF number CHE-1126177).

Author Contributions: Francisco Montagner, Danieli C. Rodrigues, Kelli Palmer, and Mihaela C. Stefan conceived the experiments; Ridwan Haseeb, Michael Lau, Max Sheah, and Gina Quiram performed the experiments. All the authors analyzed the data and wrote the manuscript.

Materials **2016**, *9*, 452

Conflicts of Interest: The authors deny any conflicts of interest. We affirm that we have no financial affiliation (e.g., employment, direct payment, stock holdings, retainers, consultantships, patent licensing arrangements, or honoraria) or involvement with any commercial organization with direct financial interest in the subject or materials discussed in this manuscript, nor have any such arrangements existed in the past three years. Any other potential conflict of interest is disclosed.

Abbreviations

The following abbreviations are used in this manuscript:

CHX	chlorhexidine
EDS	energy dispersive X-ray spectroscopy
^{1}H NMR	proton nuclear magnetic resonance
AFM	atomic force microscopy
SEM	scanning electron microscopy
ZOI	zone of inhibition
OD	optical density

References

1. Siqueira, J.F., Jr.; Rôças, I.N. Community as the unit of pathogenicity: An emerging concept as to the microbial pathogenesis of apical periodontitis. *Oral Surg. Oral Med. Oral Pathol. Oral Radiol. Endod.* **2009**, *107*, 870–878. [CrossRef] [PubMed]
2. Peters, L.B.; Wesselink, P.R. Periapical healing of endodontically treated teeth in one and two visits obturated in the presence or absence of detectable microorganisms. *Int. Endod. J.* **2002**, *35*, 660–667. [CrossRef] [PubMed]
3. Vera, J.; Siqueira, J.F., Jr.; Ricucci, D.; Loghin, S.; Fernández, N.; Flores, B.; Cruz, A.G. One-*versus* two-visit endodontic treatment of teeth with apical periodontitis: A histobacteriologic study. *J. Endod.* **2012**, *38*, 1040–1052. [CrossRef] [PubMed]
4. Agrafioti, A.; Tzimpoulas, N.E.; Kontakiotis, E.G. Influence of dentin from the root canal walls and the pulp chamber floor on the pH of intracanal medicaments. *J. Endod.* **2013**, *39*, 701–703. [CrossRef] [PubMed]
5. Haapasalo, H.K.; Sirén, E.K.; Waltimo, T.M.; Ørstavik, D.; Haapasalo, M.P. Inactivation of local root canal medicaments by dentine: An *in vitro* study. *Int. Endod. J.* **2000**, *33*, 126–131. [CrossRef] [PubMed]
6. Haapasalo, M.; Qian, W.; Portenier, I.; Waltimo, T. Effects of dentin on the antimicrobial properties of endodontic medicaments. *J. Endod.* **2007**, *33*, 917–925. [CrossRef] [PubMed]
7. Dametto, F.R.; Ferraz, C.C.R.; Gomes, B.P.F.G.; Zaia, A.A.; Teixeira, F.B.; Souza-Filho, F.J. *In vitro* assessment of the immediate and prolonged antimicrobial action of chlorhexidine gel as an endodontic irrigant against *Enterococcus faecalis. Oral. Surg. Oral. Med. Oral. Pathol. Oral. Radiol. Endod.* **2005**, *99*, 768–772. [CrossRef] [PubMed]
8. Ercan, E.; Ozekinci, T.; Atakul, F.; Gul, K. Antibacterial activity of 2% chlorhexidine gluconate and 5.25% sodium hypochlorite in infected root canal: An *in vivo* study. *J. Endod.* **2004**, *30*, 84–87. [CrossRef] [PubMed]
9. Leonardo, M.R.; Tanomaru, F.M.; Silva, L.A.; Nelson, F.P.; Bonifacio, K.C.; Ito, I.Y. *In vivo* antimicrobial activity of 2% chlorhexidine used as a root canal irrigation solution. *J. Endod.* **1999**, *25*, 167–171. [CrossRef]
10. Gomes, B.P.F.A.; Montagner, F.; Berber, V.B.; Zaia, A.A.; Ferraz, C.C.R.; de Almeida, J.F.A.; Souza-Filho, F.J. Antimicrobial action of intracanal medicaments on the external root surface. *J. Dent.* **2009**, *37*, 76–81. [CrossRef] [PubMed]
11. Murthy, S.K. Nanoparticles in modern medicine: State of the art and future challenges. *Int. J. Nanomed.* **2007**, *2*, 129–141.
12. Abbaszadegan, A.; Nabavizadeh, M.; Gholami, A.; Aleyasin, Z.S.; Dorostkar, S.; Saliminasab, M.; Ghasemi, Y.; Hemmateenejad, B.; Sharghi, H. Positively charged imidazolium-based ionic liquid-protected silver nanoparticles: A promising disinfectant in root canal treatment. *Int. Endod. J.* **2015**, *48*, 790–800. [CrossRef] [PubMed]
13. Javidi, M.; Afkhami, F.; Zarei, M.; Ghazvini, K.; Rajabi, O. Efficacy of a combined nanoparticulate/calcium hydroxide root canal medication on elimination of *Enterococcus faecalis. Aust. Endod. J* **2014**, *40*, 61–65. [CrossRef] [PubMed]

14. Barros, J.; Silva, M.G.; Rôças, I.N.; Gonçalves, L.S.; Alves, F.F.; Lopes, M.A.; Pina-Vaz, I.; Siqueira, J.F., Jr. Antibiofilm effects of endodontic sealers containing quaternary ammonium polyethylenimine nanoparticles. *J. Endod.* **2014**, *40*, 1167–1171. [CrossRef] [PubMed]
15. Kesler Shvero, D.; Abramovitz, I.; Zaltsman, N.; Perez Davidi, M.; Weiss, E.I.; Beyth, N. Towards antibacterial endodontic sealers using quaternary ammonium nanoparticles. *Int. Endod. J.* **2013**, *46*, 747–754. [CrossRef] [PubMed]
16. Álvarez, A.L.; Espinar, F.O.; Méndez, J.B. The application of microencapsulation techniques in the treatment of endodontic and periodontal diseases. *Pharmaceutics* **2011**, *3*, 538–571. [CrossRef] [PubMed]
17. Suri, S.S.; Fenniri, H.; Singh, B. Nanotechnology-based drug delivery systems. *J. Occup. Med. Toxicol.* **2007**. [CrossRef] [PubMed]
18. Shrestha, A.; Kishen, A. Antibiofilm efficacy of photosensitizer-functionalized bioactive nanoparticles on multispecies biofilm. *J. Endod.* **2014**, *40*, 1604–1610. [CrossRef] [PubMed]
19. Shrestha, S.; Diogenes, A.; Kishen, A. Temporal-controlled release of bovine serum albumin from chitosan nanoparticles: Effect on the regulation of alkaline phosphatase activity in stem cells from apical papilla. *J. Endod.* **2014**, *40*, 1349–1354. [CrossRef] [PubMed]
20. Fay, F.; Linossier, I.; Legendre, G.; Vallée-Réhel, K. Micro-encapsulation and antifouling coatings: Development of poly(lactic acid) microspheres containing bioactive molecules. *Macromol. Symp.* **2008**, *272*, 45–51. [CrossRef]
21. Gomes, B.P.; Vianna, M.E.; Zaia, A.A.; Almeida, J.F.; Souza-Filho, F.J.; Ferraz, C.C. Chlorhexidine in endodontics. *Braz. Dent. J.* **2013**, *24*, 89–102. [CrossRef] [PubMed]
22. Lboutounne, H.; Chaulet, J.F.; Ploton, C.; Falson, F.; Pirot, F. Sustained ex vivo skin antiseptic activity of chlorhexidine in poly(ε-caprolacton) nanocapsule encapsulated form and as digluconate. *J. Control. Release* **2002**, *82*, 319–334. [CrossRef]
23. Athanasiou, K.A.; Niederauer, G.G.; Agrawal, C.M. Sterilization, toxicity, biocompatibility and clinical applications of polylactic acid/polyglycolic acid copolymers. *Biomaterials* **1996**, *17*, 93–102. [CrossRef]
24. Shive, M.S.; Anderson, J.M. Biodegradation and biocompatibility of PLA and PLGA microspheres. *Adv. Drug Deliv. Rev.* **1997**, *28*, 5–24. [CrossRef] [PubMed]
25. Lopes, M.B.; Sinhoreti, M.A.; Gonini Júnior, A.; Consani, S.; Mccabe, J.F. Comparative study of tubular diameter and quantity for human and bovine dentin at different depths. *Braz. Dent. J.* **2009**, *20*, 279–283. [CrossRef] [PubMed]
26. Appel, E.A.; del Barrio, J.; Loh, X.J.; Scherman, O.A. Supramolecular polymeric hydrogels. *Chem. Soc. Rev.* **2012**, *41*, 6195–6214. [CrossRef] [PubMed]
27. Salem, A.K.; Rose, F.R.; Oreffo, R.O.; Yang, X.; Davies, M.C.; Mitchell, J.R.; Shakesheff, K.M. Porous Polymer and Cell Composites That Self-Assemble *in Situ*. *Adv. Mater.* **2003**, *15*, 210–213. [CrossRef]
28. Siqueira, J.F.; Magalhães, K.M.; Rôças, I.N. Bacterial reduction in infected root canals treated with 2.5% NaOCl as an irrigant and calcium hydroxide/camphorated paramonochlorophenol paste as an intracanal dressing. *J. Endod.* **2007**, *33*, 667–672. [CrossRef] [PubMed]
29. Upadya, M.; Shrestha, A.; Kishen, A. Role of efflux pump inhibitors on the antibiofilm efficacy of calcium hydroxide, chitosan nanoparticles, and light-activated disinfection. *J. Endod.* **2011**, *37*, 1422–1426. [CrossRef] [PubMed]
30. Greenstein, G.; Berman, C.; Jaffin, R. Chlorhexidine: An adjunct to periodontal therapy. *J. Periodontal.* **1986**, *57*, 370–377. [CrossRef] [PubMed]
31. Hugo, W.B.; Longworth, A.R. Some aspects of the mode of action of chlorhexidine. *J. Pharm. Pharmacol.* **1964**, *16*, 655–662. [CrossRef] [PubMed]
32. Böttcher, D.E.; Sehnem, N.T.; Montagner, F.; Parolo, C.C.F.; Grecca, F.S. Evaluation of the effect of *Enterococcus faecalis* biofilm on the 2% chlorhexidine substantivity: An *in vitro* study. *J. Endod.* **2015**, *41*, 1364–1370. [CrossRef] [PubMed]
33. Gomes, B.P.; Martinho, F.C.; Vianna, M.E. Comparison of 2.5% sodium hypochlorite and 2% chlorhexidine gel on oral bacterial lipopolysaccharide reduction from primarily infected root canals. *J. Endod.* **2009**, *35*, 1350–1353. [CrossRef] [PubMed]
34. Rôças, I.N.; Provenzano, J.C.; Neves, M.A.; Siqueira, J.F. Disinfecting Effects of Rotary Instrumentation with Either 2.5% Sodium Hypochlorite or 2% Chlorhexidine as the Main Irrigant: A Randomized Clinical Study. *J. Endod.* **2016**, *42*, 943–947. [PubMed]

35. Almyroudi, A.; Mackenzie, D.; McHugh, S.; Saunders, W.P. The effectiveness of various disinfectants used as endodontic intracanal medications: An *in vitro* study. *J. Endod.* **2002**, *28*, 163–167. [CrossRef] [PubMed]

36. Gomes, B.P.; Vianna, M.E.; Sena, N.T.; Zaia, A.A.; Ferraz, C.C.; de Souza Filho, F.J. *In vitro* evaluation of the antimicrobial activity of calcium hydroxide combined with chlorhexidine gel used as intracanal medicament. *Oral Surg. Oral Med. Oral Pathol. Oral Radiol. Endod.* **2006**, *102*, 544–550. [PubMed]

37. Gama, T.G.; de Oliveira, J.C.M.; Abad, E.C.; Rôças, I.N.; Siqueira, J.F., Jr. Postoperative pain following the use of two different intracanal medications. *Clin. Oral Investig.* **2008**, *12*, 325–330. [CrossRef] [PubMed]

38. Barbour, M.E.; Maddocks, S.E.; Wood, N.J.; Collins, A.M. Synthesis, characterization, and efficacy of antimicrobial chlorhexidine hexametaphosphate nanoparticles for applications in biomedical materials and consumer products. *Int. J. Nanomed.* **2013**, *8*, 3507–3519. [CrossRef] [PubMed]

39. Wood, N.J.; Maddocks, S.E.; Grady, H.J.; Collins, A.M.; Barbour, M.E. Functionalization of ethylene vinyl acetate with antimicrobial chlorhexidine hexametaphosphate nanoparticles. *Int. J. Nanomed.* **2014**, *27*, 4145–4152. [CrossRef]

40. Yue, I.C.; Poff, J.; Cortés, M.E.; Sinisterra, R.D.; Faris, C.B.; Hildgen, P.; Langer, R.; Shastri, V.P. A novel polymeric chlorhexidine delivery device for the treatment of periodontal disease. *Biomaterials* **2004**, *25*, 3743–3750. [CrossRef] [PubMed]

41. Blasi, P.; D'Souza, S.S.; Selmin, F.; DeLuca, P.P. Plasticizing effect of water on poly(lactide-co-glycolide). *J. Control. Release* **2005**, *108*, 1–9. [CrossRef] [PubMed]

42. Burdick, J.A.; Anseth, K.S. Photoencapsulation of osteoblasts in injectable RGD-modified PEG hydrogels for bone tissue engineering. *Biomaterials* **2002**, *23*, 4315–4323. [CrossRef]

43. Margelos, J.; Eliades, G.; Verdelis, C.; Palaghias, G. Interaction of calcium hydroxide with zinc oxide-eugenol type sealers: A potential clinical problem. *J. Endod.* **1997**, *23*, 43–48. [CrossRef]

44. Singh, R.P.; Ramarao, P. Accumulated polymer degradation products as effector molecules in cytotoxicity of polymeric nanoparticles. *Toxicol. Sci.* **2013**, *136*, 131–143. [CrossRef] [PubMed]

45. Cabiscol Català, E.; Tamarit Sumalla, J.; Ros Salvador, J. Oxidative stress in bacteria and protein damage by reactive oxygen species. *Int. Microbiol.* **2000**, *3*, 3–8.

46. Singh, R.P.; Ramarao, P. Cellular uptake, intracellular trafficking and cytotoxicity of silver nanoparticles. *Toxicol. Lett.* **2012**, *213*, 249–259. [CrossRef] [PubMed]

MDPI AG

St. Alban-Anlage 66

4052 Basel, Switzerland

Tel. +41 61 683 77 34

Fax +41 61 302 89 18

http://www.mdpi.com

Materials Editorial Office

E-mail: materials@mdpi.com

http://www.mdpi.com/journal/materials